Synthesis Lectures on Engineering, Science, and Technology

The focus of this series is general topics, and applications about, and for, engineers and scientists on a wide array of applications, methods and advances. Most titles cover subjects such as professional development, education, and study skills, as well as basic introductory undergraduate material and other topics appropriate for a broader and less technical audience.

Zheyi Li · Laurent Berti · Paul Leroux

Radiation Tolerant Nyquist Analog to Digital Converters

 Springer

Zheyi Li
Interuniversity Microelectronics Centre
Leuven, Belgium

KU Leuven
Geel, Belgium

Paul Leroux
KU Leuven
Geel, Belgium

Laurent Berti
Interuniversity Microelectronics Centre
Leuven, Belgium

ISSN 2690-0300 ISSN 2690-0327 (electronic)
Synthesis Lectures on Engineering, Science, and Technology
ISBN 978-3-031-95598-3 ISBN 978-3-031-95599-0 (eBook)
https://doi.org/10.1007/978-3-031-95599-0

© The Editor(s) (if applicable) and The Author(s), under exclusive license to Springer
Nature Switzerland AG 2026

This work is subject to copyright. All rights are solely and exclusively licensed by the Publisher, whether the whole or part of the material is concerned, specifically the rights of translation, reprinting, reuse of illustrations, recitation, broadcasting, reproduction on microfilms or in any other physical way, and transmission or information storage and retrieval, electronic adaptation, computer software, or by similar or dissimilar methodology now known or hereafter developed.
The use of general descriptive names, registered names, trademarks, service marks, etc. in this publication does not imply, even in the absence of a specific statement, that such names are exempt from the relevant protective laws and regulations and therefore free for general use.
The publisher, the authors and the editors are safe to assume that the advice and information in this book are believed to be true and accurate at the date of publication. Neither the publisher nor the authors or the editors give a warranty, expressed or implied, with respect to the material contained herein or for any errors or omissions that may have been made. The publisher remains neutral with regard to jurisdictional claims in published maps and institutional affiliations.

This Springer imprint is published by the registered company Springer Nature Switzerland AG
The registered company address is: Gewerbestrasse 11, 6330 Cham, Switzerland

If disposing of this product, please recycle the paper.

Preface

This book presents the process and steps taken to achieve the objective of developing a high-performance, radiation-tolerant analog-to-digital converter (ADC) for space applications. First, the principles of radiation effects in semiconductor devices are examined to provide a theoretical foundation. Then, a comprehensive technology evaluation flow and a design flow for radiation-hardened Integrated Circuits (ICs) are formulated. These flows provide detailed instructions for the design and verification of the final ADC, which is also valuable for future design and research. Subsequently, an evaluation chip, Godzilla, is designed and manufactured to evaluate the radiation tolerance of transistors in 65 nm CMOS technology. In developing the ADC, the trade-offs between power, speed, accuracy and radiation tolerance were carefully weighed. The measurement results of the ADC show exceptional performance in both power efficiency and radiation tolerance, successfully achieving the research objective. The prototype ADC achieves 70.79 dB SNDR and 80.26 dB SFDR at the Nyquist input frequency and a sampling rate of 80 MS/s. In addition, TID irradiation tests confirm that the ADC remains unaffected up to 500 krad(Si). The ADC has a limited SEE-sensitive range and also recovers quickly from SEE events.

Leuven, Belgium Zheyi Li
April 2025

Contents

1 **Introduction** 1
   1.1 The Need for High-Performance Data Converters for Space Projects 1
   1.2 Conventional and Radiation-Tolerant ADC Design and Challenge 3
   1.3 Research Goals and Objectives 4
   1.4 Book Organization 6

2 **Radiation Effects in CMOS Technology and Mitigating Techniques** 9
   2.1 Introduction 9
   2.2 Radiation Basics 10
      2.2.1 Wave-Particle Duality and Particle Energy 10
      2.2.2 Linear Energy Transfer (LET) 10
      2.2.3 Basic Radiation Effects in Materials 12
      2.2.4 Energetic (Quasi-)Particles and Interactions in Semiconductor Material 13
   2.3 Radiation Environments 15
   2.4 Space Radiation Environment 16
   2.5 Radiation Effects in Semiconductor Devices 19
      2.5.1 Dose Effects 19
      2.5.2 Single-Event Effects (SEEs) 23
      2.5.3 High Dose Rate (HDR) Effects 27
   2.6 Radiation Effects Trend in Advanced Field Effect Transistor Technologies 28
      2.6.1 Advanced Field Effect Transistor Technologies 28
      2.6.2 TID Effects Trend in Advanced Field Effect Transistor Technologies 29
      2.6.3 SEEs Trend in Advanced Field Effect Transistor Technologies 32

		2.7	Radiation Hardening Techniques	34
			2.7.1 Radiation Hardening by Process (RHBP)	35
			2.7.2 Radiation Hardening by Design (RHBD)	36
		2.8	Conclusion	39
3	**Radiation Hardened IC Design and Evaluation Flow**			**41**
		3.1	Introduction	41
		3.2	Performance Evaluation Flow for CMOS Technology to Radiation Effects	42
			3.2.1 Exploration Chip Design	43
			3.2.2 Radiation Testing	44
			3.2.3 Hardening Against Radiation Effects	46
		3.3	Design Flow for Radiation-Tolerant Library, IC and ASIC	47
			3.3.1 Design Flow	47
			3.3.2 Test Flow	49
		3.4	Conclusion	50
4	**Technology Evaulation Design Consideration and a 65-nm CMOS Technology Test Vehicle (Godzilla) for SET Evaluation**			**51**
		4.1	Introduction	51
		4.2	The Methodology of Technology Evaluation Under Radiation	51
			4.2.1 Technology Degradation Generation Circuits	52
			4.2.2 Readout	56
		4.3	65-nm CMOS Technology Test Vehicle (Godzilla) for SET Evaluation	58
			4.3.1 Position of the 65 nm CMOS Technology	58
			4.3.2 Status of the 65 nm CMOS Technology Evaluation Under Radiation	59
			4.3.3 The Scale of the Test Vehicle Godzilla	61
		4.4	Target Devices	61
			4.4.1 Target Devices Sensitive Area Calculation	61
			4.4.2 Selected Victim Devices	63
		4.5	Ionization Charge Measurement Circuits	64
			4.5.1 Main Measurement Circuits	64
			4.5.2 Charge Measurement Calibration	65
		4.6	Pulse Duration Measurement Circuits	67
			4.6.1 Main Measurement Circuits	67
			4.6.2 Pulse Measurement Calibration	69
		4.7	Test Chip and Test Setup	70
		4.8	SET Experimental Results	71
			4.8.1 Test Chips and Heavy Ion Test Conditions	71
			4.8.2 Ionization Charge Measurement Results	72

		4.8.3 Pulse Duration Measurement Results	77
	4.9	Conclusion	82
5	**ADC Fundamentals and Design Tradeoffs**		85
	5.1	Introduction	85
	5.2	ADC Principle	86
		5.2.1 Sampling	86
		5.2.2 Quantization	88
	5.3	Errors in a Non-ideal ADC	90
		5.3.1 Noise	90
		5.3.2 Non-linearity	92
	5.4	ADC Architectures	95
		5.4.1 Flash ADCs	96
		5.4.2 Pipeline ADCs	96
		5.4.3 SAR ADCs	97
		5.4.4 Sigma Delta ($\Sigma\Delta$) ADCs	97
		5.4.5 Hybrid ADCs	98
	5.5	ADC Design Tradeoff	98
	5.6	Radiation-Hardened ADC Design Tradeoffs	100
		5.6.1 Total Ionizing Dose Effects	102
		5.6.2 Single Event Effects	104
		5.6.3 ADC Structure Selection Consideration	107
	5.7	Conclusion	109
6	**Radiation Hardened Pipelined-SAR ADC Architectural Modeling and Design Considerations**		111
	6.1	Introduction	111
	6.2	SAR-Assisted Pipeline Structure	112
	6.3	Bit Division Modeling in SAR-Assisted Pipeline ADC	113
	6.4	CDAC Switching Power Analysis	120
	6.5	CDAC Capacitor Mismatch and Calibration	123
		6.5.1 CDAC Capacitor Mismatch Limitation	123
		6.5.2 Previous CDAC Calibration Methods	125
		6.5.3 Proposed Calibration Principle	126
	6.6	Radiation Tolerance of the Sub-blocks in SAR-Assisted Pipeline ADC	129
	6.7	Conclusion	132
7	**13-Bit High-Performance Radiation-Tolerant ADC**		133
	7.1	Introduction	133
	7.2	ADC Top-Level Architecture Selection	134
	7.3	Proposed ADC Structure with Efficiency Improvement	135
	7.4	Design Details of the Key ADC Sub-blocks	137

		7.4.1	Shared Residue Amplifier	137
		7.4.2	Clock Generation	145
		7.4.3	Comparator and Time-Out Protection	146
		7.4.4	Digital Circuits and Transistors	149
		7.4.5	Layout of the Coarse Stage ADC	149
	7.5	Experimental Results		151
		7.5.1	Electronic Measurement Results	152
		7.5.2	Irradiation Test Results	154
	7.6	Performance Summarize		159
	7.7	Conclusion		159
8	**Conclusions and Future Research**			161
	8.1	Introduction		161
	8.2	General Conclusions		161
	8.3	Research Novelties		162
	8.4	Future Research		163
		8.4.1	Further Investigation of 65 nm CMOS Technology	163
		8.4.2	Improvement of the Prototype ADC	164
		8.4.3	Future ADC Designs	165
9	**Research Valorization Feasibility Study**			167
	9.1	Introduction		167
	9.2	Present State of Space Market		168
	9.3	Offer to the Market		169
		9.3.1	AT-AD1.0 High Performance Radiation-Hardened ADC	169
		9.3.2	Products and Services Roadmap	170
		9.3.3	Business Model	171
	9.4	Conclusion		171
References				173

Acronyms

ADC	Analog Design Kit
ADC	Analog-to-Digital Converter
ASET	Analog Single Event Transient
ASIC	Application-Specific Integrated Circuit
ATLAS	A Toroidal LHC Apparatus
AU	Astronomical Unit
AZ	Auto-Zeroing
BOX	Buried Oxide
C-ADC	Coarse-stage ADC
CBW	Conventional Binary-Weighted
CCB	Correction Capacitor Bank
CDAC	Capacitor Digital-to-Analog Converter
CERN	Conseil Européen pour la Recherche Nucléaire
CME	Coronal Mass Ejection
CMOS	Complementary Metal-Oxide Semiconductor
COTS	Commercial-Off-The-Shelf
DAC	Digital-to-Analog Converter
DARE	Design Against Radiation Effects
DD	Displacement Damage
DFF	D-type Flip-Flops
DFT	Discrete Fourier Transform
DICE	Dual Interlocked Storage Cell
DMR	Double Module Redundancy
DNL	Differential Non-Linearity
DNW	Deep N-Well
DRAM	Dynamic Random Access Memory
DRC	Design Rule Check
DRS	Detection Reference Setting

DSET	Digital Single Event Transient
DUT	Device Under Test
EDA	Electronic Design Automation
EDAC	Error Detection And Correction
ELT	Enclosed Layout Transistor
ENOB	Effective Number of Bits
ESA	European Space Agency
ESD	Electrostatic Discharge
F-ADC	Fine-stage ADC
FD-SOI	Fully Depleted Silicon On Insulator
FET	Field Effect Transistor
FoM	Figures of Merit
FPGA	Field Programmable Gate Arrays
GAA	Gate-All-Around
GCR	Galactic Cosmic Ray
HDR	High Dose Rate
HIF	Heavy Ion Facility
IC	Integrated Circuit
IEEE	Institute of Electrical and Electronics Engineers
IMEC	Interuniversity Microelectronics Center
INL	Integral Non-Linearity
IO	Input/Output
ISSCC	International Solid-State Circuits Conference
LDO	Low-Dropout Regulator
LEO	Low Earth Orbit
LET	Linear Energy Transfer
LHC	Large Hadron Collider
LOCOS	Local Oxidation of Silicon
LSB	Least Significant Bit
LVDS	Low-Voltage Differential Signaling
MBU	Multiple Bit Upset
MCS	Merged Capacitor Switching
MDC	Memory Delay Cell
MDL	Memory Delay Line
MIM	Metal-Insulator-Metal
MOM	Metal-Oxide-Metal
MSB	Most Significant Bit
MSE	Mismatch Sign Extraction
NASA	National Aeronautics and Space Administration
PCB	Printed Circuit Board
PDK	Process Design Kit

PD-SOI	Partially Depleted Silicon On Insulator
PIPB	Propagation-Induced Pulse Broadening
PVT	Process, Voltage, and Temperature
RA	Residue Amplifier
RHBD	Radiation Hardening by Design
RHBP	Radiation Hardening by Process
RISCE	Radiation-Induced Short-Channel Effect
S&H	Sample-and-Hold
SADA	Solar Array Drive Assembly
SAR	Successive Approximation Register
SBU	Single Bit Upset
SCE	Short Channel Effect
SEB	Single Event Burnout
SEE	Single Event Effect
SEFI	Single-Event Functional Interrupt
SEGR	Single Event Gate Rupture
SEL	Single Event Latchup
Semi-TI	Semi-Time-Interleaved
SEP	Solar Energetic Particles
SET	Single Event Transient
SEU	Single Event Upset
SNDR	Signal-to-Noise-Distortion Ratio
SOI	Silicon On Insulator
SQNR	Signal-to-Quantization-Noise Ratio
SRAM	Static Random Access Memory
STI	Shallow-Trench Isolation
TCAD	Technology Computer-Aided Design
TID	Total Ionizing Dose
TMR	Triple Module Redundancy
TPA	Two-Photon Absorption
UTBB	Ultra-Thin-Body and BOX
VLSI	Symposium on VLSI Technology and Circuits

Introduction

Abstract

This chapter introduces the overall theme of the book, focusing on the need for high-performance Analog-to-Digital Converters (ADCs) in space applications. It outlines the limitations of conventional ADC designs in radiation environments and presents the key challenges in developing radiation-tolerant solutions. By highlighting the gap between performance and robustness, the chapter sets the foundation for the research presented in the following chapters, where advanced design techniques for reliable ADC operation under radiation conditions are explored.

1.1 The Need for High-Performance Data Converters for Space Projects

In recent years, the continuous increase in investment by various countries in the exploration and application of space has led to a growing demand for high-performance electronics for space applications. As of February 2023, Starlink consists of 3,580 mass-produced small satellites in Low Earth Orbit (LEO) that communicate with the designated transceivers on the ground. In total, the deployment of almost 12,000 satellites is planned, with a possible later expansion to 42,000 [1]. Starting in 2022 and throughout the decade, the Artemis projects of National Aeronautics and Space Administration (NASA) and European Space Agency (ESA) will send a series of scientific instruments and technological demonstrations to the lunar surface through commercial lunar payload deliveries [2]. All of these space projects, involving satellites, rovers or deep probes, require sophisticated electronic systems capable of collecting, processing and transmitting large amounts of data. Therefore, these systems must be robust in order to survive in the radiation environment of space.

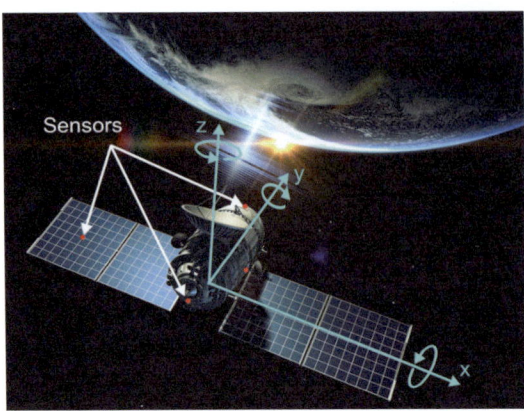

Fig. 1.1 Tilt sensors on a satellite for solar panel rotation

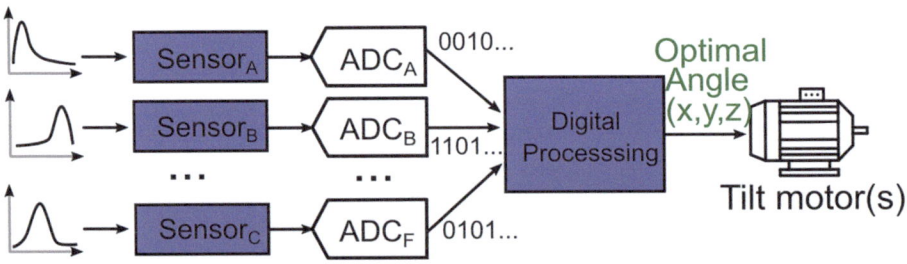

Fig. 1.2 Signal processing chain in a solar array orientation system in satellites

Among the critical components within electronic systems, high-performance data converters, such as ADCs and Digital-to-Analog Converters (DACs), play a central role. Data converters bridge the gap between continuous analog signals and discrete digital signals used in digital systems that can only process 0 and 1 [3]. As an example, Figs. 1.1 and 1.2 show a solution for a solar array orientation system in satellites. The generation of electrical power is essential to ensure the survival of the space system. When a satellite is powered by the electricity generated by solar arrays, the ability to align them with the sun is an essential requirement for the system [4]. The orientation of the solar cells can be achieved with a Solar Array Drive Assembly (SADA) system, with an appropriate rotation of the spacecraft or the solar panels, or with a combination of both [4, 5]. Several sensors distributed throughout the satellite monitor the information on velocity, three-axis attitude, light intensity, etc. and deliver them to the SADA system. As the sensors can only provide a continuous time and amplitude signal, ADCs are required to convert the signal into time and amplitude discrete binary bits that the on-board computer can process. The digital algorithm then calculates the

1.2 Conventional and Radiation-Tolerant ADC Design and Challenge

optimum alignment angle of the solar modules and gives the motors the command to perform the rotation. In this SADA system, the ADCs must be accurate, consume little power and work reliably throughout the entire space mission.

Conventional ADC design challenge: Based on different application scenarios, the design of ADC focuses on different metrics. In general, the conventional ADC performance is limited by the three-dimensional tradeoff: Power, speed and accuracy (shown in Fig. 1.3) [3, 6, 7]. To improve accuracy, higher supply/reference voltages are required to extend the dynamic range. A higher bias current can be used to reduce noise, and larger components improve matching and linearity. All of them cost more power consumption and reduce the ADC speed. On the other hand, fast conversion speed requires a fast sample and hold function, logic, comparator and amplifier, which easily drives up power consumption or reduces accuracy. The overall performance and efficiency of ADCs are usually evaluated by the Figures of Merit (FoM). The specific formula used to calculate the ADC FoM may vary depending on the application. Commonly used FoMs for ADCs include the Walden Figure of Merit (FoM_W) [8], which is represented in the Eq. (1.1), and the Schreier Figure of Merit (FoM_S) [9], which is represented in the Eq. (1.2). Both FoMs reveal the above-mentioned trade-off.

$$FoM_W = \frac{P}{f_s \times 2^{ENOB}} [fJ/conversionstep] \qquad (1.1)$$

$$FoM_S = SNDR + 10\log\left(\frac{f_s}{2P}\right) [dB] \qquad (1.2)$$

In the equations above, P stands for the total ADC power consumption and f_s stands for the sampling frequency. ENOB and SNDR are abbreviations for the Effective Number of Bits and Signal-to-Noise-Distortion Ratio respectively. Both are related to the accuracy of the ADC, which is higher the better it is. Although the conventional ADC trade-off is difficult to eliminate, it is largely influenced by different design levels: system level, architecture level, circuit level and technology level. ADC performance can be significantly improved by design techniques such as time interleaving (system level), pipelined structure (architecture level), dynamic power control (circuit level), and backgate voltage control in Fully Depleted Silicon On Insulator (FD-SOI) technologies (technology level).

Radiation-tolerant ADC design challenge: Radiation effects in space, such as Single Event Effects (SEEs) and Total Ionizing Dose (TID) effects, can cause temporary or permanent malfunctions and performance degradation in electronic devices and also in ADCs [10–12]. These radiation effects are explored in Chap. 2. Due to the harsh conditions in the

Fig. 1.3 Design trade-offs in radiation-tolerant ADCs in space projects

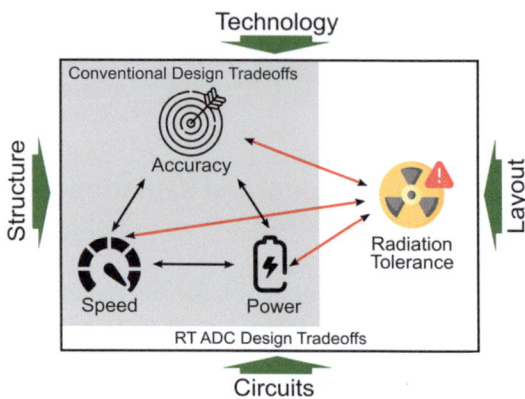

space environment, radiation tolerance has proven to be a critical factor when considering ADCs for space projects, as shown in Fig. 1.3. To ensure the reliable operation of electronic systems in space missions, robustness takes precedence over other design aspects. Therefore, various design techniques, such as redundancy and cold/hot spare, are used to ensure the functionality and performance stability of ADCs throughout the lifetime of the mission. As a result, radiation-tolerant ADCs often exhibit lower efficiency (higher FoM_W and lower FoM_S) compared to conventional ADCs when speed and accuracy are the same metrics. Figure 1.4 shows the FoM_W of published Nyquist ADCs since 2010 [13]. All these ADCs are from IEEE ISSCC and VLSI and were implemented with technology advanced or equal to 90 nm. In addition, four radiation-tolerant ADCs are marked in the figure. Two of them are from published works [14, 15], and the other two are from the market [16, 17]. It can be seen that radiation-tolerant ADCs have an order of magnitude lower efficiency compared to conventional ADCs. Therefore, improving the efficiency of radiation-tolerant ADCs is favored in space projects, as power generation and storage on board spacecraft are limited. Higher power consumption directly leads to a larger area for the solar cells and higher weight of the batteries, which in turn leads to higher launch weight, higher overall project cost, and higher project risk.

1.3 Research Goals and Objectives

The main and ultimate goal of this research is to develop a high-performance, radiation-tolerant ADC using 65 nm bulk Complementary Metal-Oxide Semiconductor (CMOS) technology. The sampling frequency needs to be higher than 50 MS/s, and the accuracy must be higher than 12 bits. In addition to radiation tolerance, this ADC should also have comparable efficiency (FoM) compared to the conventional ADCs in similar metrics. The specific objectives and tasks are described below:

1.3 Research Goals and Objectives

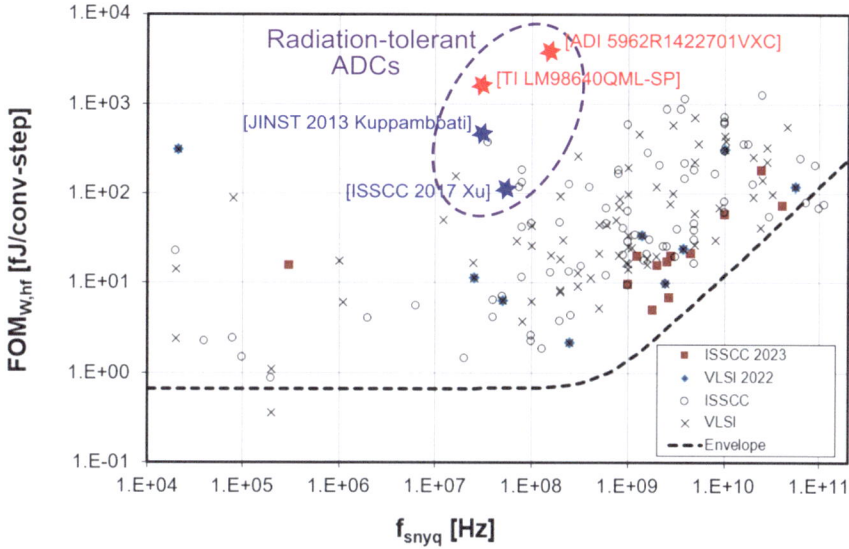

Fig. 1.4 ADC Walden FoM versus speed, with following conditions: publish year later than 2010, technology is more advanced or equal to 90 nm and Nyquist ADCs (*Source* Murmann [13])

- **Investigate the radiation effects on CMOS technology and radiation hardening techniques**
  At the beginning of the research, the effects of radiation, such as SEEs and TID effects, on the CMOS technology were investigated. The impact of scaling the technology on these radiation effects was then investigated. This finally allows us to explore the potential radiation hardening techniques at different design levels, such as technology level, circuit level, and system level, and determine the optimal hardening techniques for the subsequent design.
- **Conclude radiation-tolerant IC, ASIC design and evaluation flow**
  In the second period, the aim was to complete the necessary steps to develop a radiation-tolerant Integrate Circuit (IC) or Application-Specific Integrated Circuit (ASIC) from scratch. This includes the technology assessment process to understand the impact of radiation on the technology. After the technology assessment, additional design and test procedures are added to the traditional IC and ASIC design flow.
- **Evaluate 65 nm CMOS technology under radiation effects**
  As the final ADC design is based on 65 nm CMOS technology, this technology needs to be well investigated and understood. The advantages and disadvantages of this technology need to be analyzed for both conventional and radiation-hardened designs. A test chip was developed to evaluate the 65 nm CMOS technology to analyze the performance under radiation. Based on this study, the optimal radiation hardening strategy for a chip in 65 nm CMOS technology is determined.

- **Design a high-performance radiation-tolerant ADC in 65 nm CMOS technology**
  In this phase, the goal is to design an ADC fulfilling the target metrics. Different types of ADCs are first compared and modeled at the system level to find the optimal structure. Next, the radiation effects on all sub-blocks in the selected ADC structure are analyzed to find suitable radiation-hardening techniques and methods. Then, the transistor-level design is performed, including simulations of the radiation effects at the block and system levels. Subsequently, the layout is drawn, and a test plan is designed to determine both the nominal performance of the ADC and its behavior under radiation. Finally, comprehensive ADC characterization and testing are performed.
- **Conclude this research and provide recommendations for future research**
  Here, the findings and insights on the effects of radiation on 65 nm CMOS technology and the high-performance radiation-tolerant ADC design are summarized. On this basis, suggestions for improvements are made and modifications are proposed to further develop the ADC prototype for a future product. Last but not least, possible avenues for further research in the field of radiation-tolerant electronics and for the exploration of new technologies for space applications are identified.

1.4 Book Organization

This book consists of nine chapters, including the present chapter. Below you will find a brief overview of the individual chapters:

- **Chapter** 2 introduces the common sources of radiation in the space environment and their impact on modern CMOS technology and circuits. The role of technology scaling in these radiation effects is discussed. The typical solutions for radiation hardening are also explored.
- **Chapter** 3 introduces the technology assessment process to evaluate the response of a technology to radiation effects. The design flow for radiation-hardened ASIC and IC is then presented. The flow is based on the conventional design flow with additional steps and checkpoints.
- **Chapter** 4 first introduces the 65 nm CMOS technology and its position in the CMOS technology roadmap. It then discusses the advantages and disadvantages of 65 nm CMOS technology when exposed to different radiation sources. By identifying the primary radiation effects specific to this technology, a dedicated test vehicle is developed to evaluate the response of 65 nm CMOS technology under these radiation effects. This chapter covers the details of the test campaign and the measurement results of the test vehicle for the 65 nm CMOS technology.

1.4 Book Organization

- **Chapter** 5 explain the principles of the analog-to-digital converters. Various ADC structures are presented and compared. The trade-offs in the development of conventional ADCs and radiation-tolerant ADCs are also discussed in this chapter.
- **Chapter** 6 presents the detailed analysis of the pipelined successive approximation register (pipelined-SAR) ADCs from the architecture level to the block level. In this chapter, the critical features of the pipelined-SAR ADC such as bit division and mismatch compensation are discussed to achieve an optimal power-efficient design. It is worth mentioning that in addition to the conventional design considerations, the radiation effects on these sub-blocks are also considered and simulated during the design process.
- **Chapter** 7 presents the details of the proposed radiation-tolerant ADC design. It also presents the ADC characterization results, which include both the conventional performance results and the radiation results.
- **Chapter** 8 summarizes the overview and draws the conclusion of this work. The chapter concludes with the most important highlights and insights gained during the research journey. In addition, this chapter provides suggestions and recommendations for future research directions to inspire and guide further progress in this field.
- **Chapter** 9 presents a possible valorization plan based on the foundation of a spin-off company whose first product is the radiation-tolerant high-performance ADC developed in this research. The ultimate vision of the spin-off is to build a comprehensive portfolio of analog/mixed-signal solutions and advanced ASIC design services, making the spin-off a major player for suppliers of microelectronic components in extreme environments.

Radiation Effects in CMOS Technology and Mitigating Techniques

Abstract

This chapter gives a brief introduction to the physical mechanisms that cause radiation effects. Two main effects on semiconductor devices—Total Ionizing Dose (TID) effects and Single Event Effects (SEEs)—are examined. When reviewing the radiation effects in different technology nodes, it was found that the planar transistors in scaled technologies are more tolerant to TID effects but more vulnerable to SEEs. Therefore, in recent years, radiation hardening has focused more on SEEs in technologies below or equal to 90 nm. The following chapter and design also put more emphasis on SEEs. Last but not least, mitigation techniques are presented at different design levels. Some of the techniques are applied to the radiation-hardened ADC design in the later chapters.

2.1 Introduction

In this chapter, we first discuss the fundamental principles underlying radiation phenomena and provide an insight into their physical origins. We then discuss various radiation environments in space, including galactic cosmic rays, solar radiation, and the radiation belts. In this context, we highlight three dominant radiation-induced effects on semiconductor devices: Dose Effects, Single-Event Effects, and High Dose Rate (HDR) effects. A comprehensive exploration of the physical principle of these effects is discussed together with the degradation of the electronic properties of CMOS.

The scaling of technology plays an important role in the development of modern electronics. It also has an impact on the radiation effects of semiconductor devices. As semiconductor technology continues to scale, smaller feature sizes and higher integration densities have made modern CMOS devices more susceptible to some effects and more resistant to others. Various CMOS technologies with different structures and process nodes and their effects on radiation are presented. Finally, different techniques and approaches to mitigate the harmful effects of radiation are examined.

This chapter is a valuable resource that provides insight into the radiation challenges facing modern CMOS technology in space applications. By examining radiation sources, hardening solutions, and technology scaling effects, we aim to provide a basic understanding of the complexities associated with developing radiation-hardened microelectronics for a wide range of space missions and applications.

2.2 Radiation Basics

Before diving into the intricacies of the radiation environment, it is important to develop a foundational understanding of radiation concepts. This knowledge provides the necessary framework to understand the effects of radiation on semiconductor devices, which we will discuss later in this chapter.

2.2.1 Wave-Particle Duality and Particle Energy

Radiation is the transport of energy from one place to another. The carriers of energy can be photons, electrons, muons or nucleons (neutrons or protons) [12]. Each particle can exhibit both particle-like and wave-like behavior, which is called Wave-Particle Duality [18]. This duality can be characterized by the following equation:

$$\lambda = \frac{h}{p} = \frac{h}{mv} = \frac{h}{\sqrt{2mE_k}} \tag{2.1}$$

where h is Plank's constant, p is the momentum of the particle, m is its mass, v is its velocity and E_k is its kinetic energy. It is known that the velocity and momentum increase with increasing energy of the particle. On the other hand, the wavelength of the particle becomes shorter as its energy increases.

2.2.2 Linear Energy Transfer (LET)

The relationship between the particle energy and the deposited energy to the target mass does not exhibit a linear function. A higher particle energy can lead to a lower deposited energy. One of the most common terms used to characterize the amount of energy transferred per unit length is Linear Energy Transfer (LET) [19]. LET is not constant, but depends on the type of particles, the energy and the target material. In addition, the LET is highly non-linear depending on the particle energy and is usually given in units of megaelectronvolt-square centimeters per milligram (MeV·cm^2/mg). Figure 2.1 shows the simulated LET (red) and

2.2 Radiation Basics

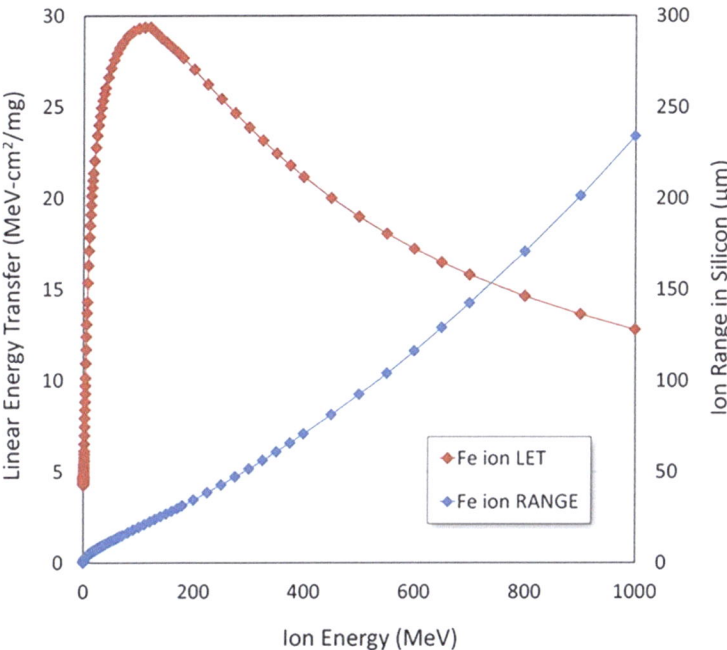

Fig. 2.1 The simulated LET (red) and range (blue) of an iron ion in a silicon target as a function of the ion's energy (*Image source* SRIM)

the range (blue) for an iron ion in silicon as a function of its energy [20]. It can be seen that the higher the ion energy, the deeper the ion can penetrate. The range of the ions in silicon increases almost linearly with the energy of the ions. However, the LET initially increases and then decreases as the ion energy increases. The peak value of the LET occurs at an energy of about 150 MeV. This nonlinear property indicates that insufficient shielding does not help to reduce the LET. On the contrary, it can increase the LET and cause more radiation-induced charges. It has been reported that shielding does not completely block the alpha particles but stops them closer to the active regions, resulting in higher LET and worse soft errors in Dynamic Random Access Memory (DRAM) [21].

The LET is also influenced by the angle of incidence of the ion strike. As shown in Fig. 2.2, the active regions of the semiconductor devices are the most charge-sensitive, and these regions are limited to thin surface layers. Therefore, when the angle of incidence of the ion is greater than 0, more ionized charges are generated in the active regions. The same LET becomes more effective in generating ionized charges as the energy particle travels a longer distance and more energy is lost in the active region. Equation (2.2) expresses the concept of the effective LET_{eff}:

$$LET_{eff} = \frac{LET}{\cos\theta} \qquad (2.2)$$

Fig. 2.2 Angle effects in the effective LET

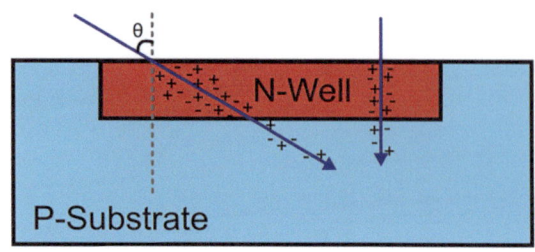

Fig. 2.3 The simulated effective LET versus iron ion energy and incidence angle (*Image source* SRIM)

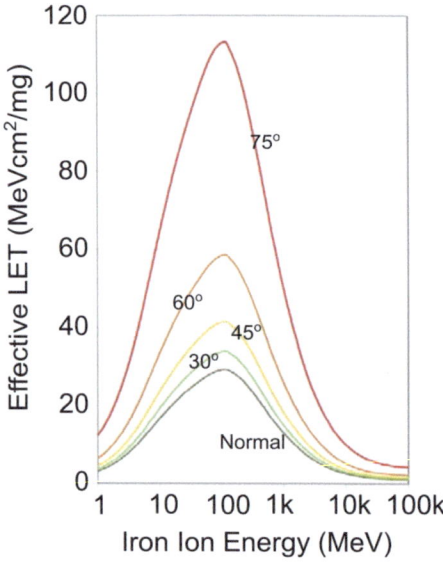

Figure 2.3 shows the simulated effective LET of iron in silicon as a function of the ion energy [20]. Each curve represents a different angle of incidence. The figure shows that the angle of incidence strongly influences the effective LET when the energy of the iron ion is around 100 MeV. The effective LET reaches its maximum value when the angle of incidence is 75°. It is worth noting that LET_{eff} is an engineering approximation and may not be accurate in some advanced semiconductor technology nodes.

2.2.3 Basic Radiation Effects in Materials

In this subsection, the primary radiation effects resulting from the interaction of radiation with electronic materials are briefly discussed. Two main primary effects are ionization effects and displacement damage (DD) effects.

Ionization effects occur when one (or more) electrons in the valence band are excited across the band gap into a conduction band state. This can be a direct result of interaction with a high-energy charged particle. In a very short time (of the order of a picosecond), the excited electron in the conduction band and the hole left in the valence band lose their excess kinetic energy through lattice scattering and are thermalized in their energy, falling to the vicinity of the conduction and valence band edges, respectively. Except for some of the electron/hole pairs that recombine, the rest are free to diffuse and drift away from their point of origin. As a result, electric current is generated when an electric field is present [22]. The most critical radiation-induced effects in modern semiconductor devices, TID effects and SEEs, are all due to ionization effects.

Displacement Damage effects refers to the physical disruption and displacement of atoms within a material caused by high-energy particles such as ions or neutrons. These high-energy particles collide with the atoms in the material, releasing enough energy to displace them from their original lattice position. This changes the atomic structure of the material and creates defects or vacancies in the crystal lattice [22, 23]. Further explanations of semiconductor material (e.g. Si) can be found in the next section.

2.2.4 Energetic (Quasi-)Particles and Interactions in Semiconductor Material

Radiation is mainly caused by electromagnetic waves and various energetic particles. Electromagnetic waves are usually defined by frequency, wavelength and photon energy. The electromagnetic spectrum is shown in Fig. 2.4 [24]. Since the electromagnetic interference radiation effects with a frequency of less than 10^{12} Hz are effectively attenuated by the standardized design, layout and packaging [12], they will not be discussed further. On the other hand, the primary radiation sources associated with modern semiconductor technology come from energized particles such as photons, electrons, nucleons (neutrons and protons) and ionized atoms [25, 26].

- **Photons** are fundamental quasiparticles of light and electromagnetic radiation. They are regarded as carriers of electromagnetic energy and have properties of both particles and waves. Photons have no rest mass and move at the speed of light in a vacuum [18, 27]. Photons with high energy, such as X-rays and gamma photons, can easily penetrate packaging materials and affect the semiconductor device through ionization. If the photon energy is high enough, photoelectric effects occur, creating electron-hole pairs. The photoelectric effect is the predominant phenomenon when photons (photon energy from 10 to 100 keV) interact with silicon [12, 28]. At higher photon energies, Compton scattering becomes the dominant effect. The photon loses part of its energy in a collision with a single electron. The scattering reaction produces a free recoil electron and a "scattered" photon, which is deflected in a different direction with less energy. At even higher photon energies,

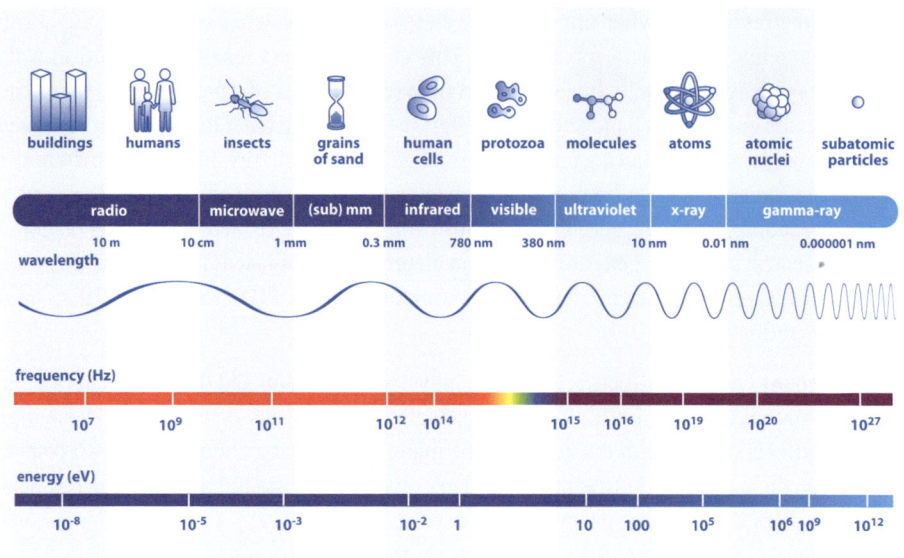

Fig. 2.4 Electromagnetic spectrum shows various properties across the range of frequencies and wavelengths (*Image source* ESA)

pair production starts and eventually becomes the dominant energy loss mechanism for high-energy gamma rays [12].

- **Electrons** are negatively charged particles. Therefore, the interaction occurs through the Coulomb force of orbital electrons and nuclei in the target material. The result of each interaction is always a redirected electron with or without the emission of a photon [28]. There are three primary mechanisms for electron interactions. They are the elastic electron-electron interaction, the inelastic electron-electron interaction and the inelastic electron-nucleus scattering[2] [29]. These interactions can lead to the deposition of energy, the creation of secondary particles (such as protons) and a change in the electron's path. Thanks to the packages, only electrons with a kinetic energy of more than 300 keV can penetrate the package and reach the die. However, in space environments, especially in radiation belts, the electron flux is high and the electrons have an energy range of 0.1 to 10 MeV. Therefore, the majority of electrons in such space environments will easily penetrate the package and cause ionizing effects [12].
- **Nucleons** comprise the protons and neutrons which build the nuclei. Neutrons and protons have almost the same weight. However, neutrons are electrically neutral, while protons are positively charged. Neutron-induced defects (e.g. displacement damage) in semiconductor

---

[2] Elastic scattering refers to a process in which the kinetic energy of the colliding particles (in this case, electrons) is conserved before and after the interaction.

devices lead to localized alterations in the electrical properties of the semiconductor. DD occurs when these effects accumulate over a long period of time. In addition, the neutron-induced recoil nucleus causes ionization on its journey and can therefore potentially cause SEEs [12]. Protons can ionize materials directly and are the primary radiation in space. Since a significant fraction of protons in space have enough energy to penetrate shielding and packaging, protons are one of the sources of SEEs. If the proton fluence is also high, TID effects and displacement damage can also occur. Protons also tend to cause significant secondary ionization after nuclear interactions.

- **Ions** are usually atoms that have lost part or all of their electrons. Energetic ions are therefore positively charged and move with high kinetic energy. In contrast to the previous particles, energetic ions deposit high-density energy (it can be hundreds of femto-coulombs per micrometer), which leads to the formation of local, filamentary, cylindrical distributions of high-density ionization charges along their path, also known as charge funnel [30, 31]. As a result, heavy ion events are the main cause of SEEs, such as turning on the bipolar mechanisms and causing Single-Event Latchup (SEL). Similar to protons, heavy ions have enough energy to easily penetrate shielding and package, but heavy ions rarely occur at high fluence.

2.3 Radiation Environments

Different criteria can classify different types of radiation environments. According to [27], different radiation environments can generally be classified into the following types (see Fig. 2.5):

- **Space radiation environment**
  Space radiation mainly includes high-energy particles such as protons, electrons and heavy ions that originate from the sun, galactic cosmic rays and the radiation belts around the planets [27, 32].
- **High-energy physics radiation environment**
  In high-energy physics experiments such as ATLAS (A Toroidal LHC Apparatus), the Large Hadron Collider (LHC) at Conseil Européen pour la Recherche Nucléaire, which is also well known as CERN [33, 34] electron or proton beams are accelerated to hundreds of GeV energy levels in order to study elementary particles with short lifetimes, such as bosons, muons and quarks.
- **Nuclear radiation environment**
  The nuclear reactors in nuclear power plants can create a mixed environment with high dose and high radiation flux, including gamma rays, X-rays and neutrons [27].

Fig. 2.5 Radiation in different environments

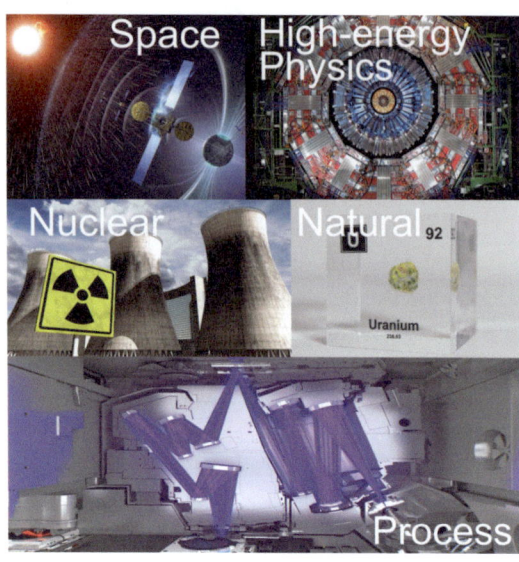

- **Natural radiation environment**
  The materials used for the packaging and soldering process may contain radioactive impurities such as uranium and thorium and their daughter products. These natural radioactive impurities produce alpha particles [27].
- **Processing-induced radiation environment**
  Last but not least, radiation damage can occur during the manufacture of modern semiconductor [35]. For example, through ion implantation, X-ray lithography and plasma-enhanced chemical vapor deposition. While radiation damage caused by the semiconductor process is relatively low compared to other radiation environments, the vulnerability of advanced technology nodes increases as the technology scales.

Since this research is primarily focused on ADCs for space applications, the main scope of the study focuses on the radiation environment in space. Therefore, other radiation environments are not specifically addressed in this study. Further references on the radiation environments can be found in [36–39].

2.4 Space Radiation Environment

The radiation environment in space originates from three main sources: the sources outside the solar system (e.g. galactic cosmic rays), the sun (e.g. solar flares, coronal mass ejection and solar wind) and the radiation belts of the planets (e.g. the Earth's radiation belts, also called Van Allen belts) [32, 40].

2.4 Space Radiation Environment

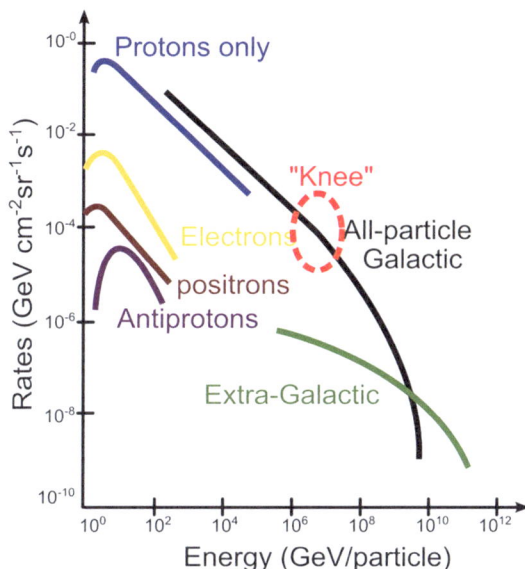

Fig. 2.6 Spectrum of cosmic rays at the earth

Galactic Cosmic Rays (GCRs) are mainly charged particles that contribute to the energy density in the galaxy of about 1 eVcm^{-3} [41]. GCRs contain 89% ionized hydrogen (protons) and 9% ionized helium (alpha particles) as well as 2% heavier ions and electrons [12, 41]. Figure 2.6 shows the different flux of GCRs as a function of particle energy. When the particle energies are below 30 GeV, the spectral shape bends downwards as they are deflected by the heliosphere, making it more difficult for the particles to reach the inner solar system. Beyond 30 GeV particle energy, the GCR flux decreases continuously with increasing particle energy. At 10^{15} eV particle energy, there is a significant change in the spectral slope from -2.7 to -3.7. When the energy exceeds the spectral knee, the chemical composition of GCRs at high energy becomes more dominated by heavy nuclei [42]. Recent measurements show an interesting structure in the spectrum and composition of GCRs up to 10^{18} eV [43].

Solar radiation is the most intense source of radiation in the solar system and comes from solar activity. Solar activity consists of the solar wind, solar flares and Coronal Mass Ejections (CMEs). The solar wind is produced by the high-energy particles that escape solar gravity due to the high temperature in the solar corona. It contains high-energy photons, electrons, protons, helium ions and a small number of heavier ions. Solar flares are often associated with CMEs. During solar flares, high magnetic energy is released, which leads to the heating and acceleration of Solar Energetic Particles (SEPs), such as electrons, protons and heavier nuclei, to high kinetic energies. Figure 2.7 shows the representative proton energy spectra at 1 Astronomical Unit (AU) [12]. It can be seen that the proton energy of the SEP events ranges from 1 MeV to 1 GeV.

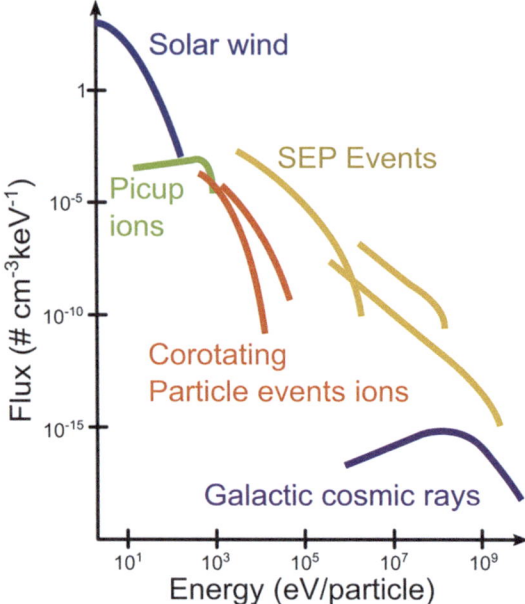

Fig. 2.7 Proton flux as a function of proton energy for solar wind, SEPs and GCRs distributions

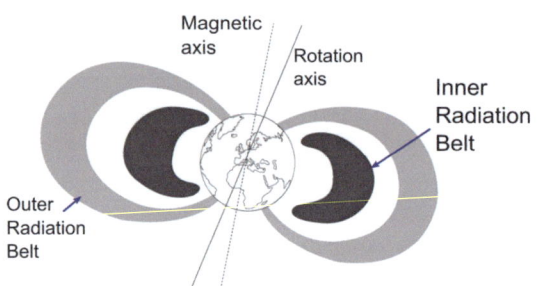

Fig. 2.8 Van Allen radiation belts (*Image source* ESA)

Radiation belts are formed due to the planet's magnetic field (magnetosphere), which can trap particles before they can penetrate the atmosphere. The Earth's radiation belts, also known as the Van Allen belts, were discovered in 1958 [40]. The Earth's magnetic field traps the protons and electrons, creating doughnut-shaped regions with concentrated charged particles (see Fig. 2.8). The belts are thicker at the equator, where the magnetic field is strong, and become thinner at higher and lower latitudes. The inner belt ranges from an altitude of 1200 to 6000 km and contains a high concentration of electrons with kinetic energies of 1~5 MeV and protons with kinetic energies of 10~100 MeV [44]. The particle composition of the outer belt varies with solar activity.

Table 2.1 summarizes the maximum particle energy in different radiation environments. It shows that shielding is not effective for many radiation environments. Therefore, radiation hardening of semiconductor devices is crucial for space applications.

2.5 Radiation Effects in Semiconductor Devices

Table 2.1 Maximum particle energy in space radiation environments

Particle type	Maximum energy levels
Trapped electrons	10s of MeV
Trapped protons	100s of MeV
Heavy ions	100s of MeV
Solar protons	GeV
Solar heavy ions	GeV
Galactic cosmic rays	>TeV

2.5 Radiation Effects in Semiconductor Devices

The radiation effects in semiconductor devices can be subdivided according to the time scale and can be listed in terms of duration from long to short:

- Dose Effects are the accumulative parametric shifts of semiconductor devices over time due to chronic exposure to radiation. The duration can be months to years and leads to performance degradation and eventually malfunction.
- SEEs are instantaneous disturbances caused by the impact of a single particle. These effects can last from femtoseconds to microseconds, depending on the circumstances in the circuit and LET. The SEEs can cause temporary failures or permanent damage to circuits and systems.
- Dose-Rate Effects happen when the semiconductor devices are exposed to extremely HDRs of radiation over a short period of time.

2.5.1 Dose Effects

The dose effects contain two main categories: TID and DD effects. Both types of effects are accumulative damage due to long-term exposure to radiation. However, the damage locations and mechanisms are different.

Total Ionizing Dose (TID) effects occur when the MOSFETs are exposed to high-energy ionizing radiation and electron-hole pairs are generated in the insulator materials.[3] In semiconductor materials such as silicon, the electron-hole pairs move and disperse in a short time due to drift and diffusion. However, in the insulator used in MOSFETs, such as the SiO_2 gate, the mobility of the electrons is much higher than that of the holes when an electric field is applied. Then the electrons will quickly removed from the oxide by the drift

---

[3] An energy of about 17 eV is required for the generation of a single electron-hole pair in silicon dioxide (SiO_2) [12].

Fig. 2.9 Threshold voltage shift due to oxide trapped charge

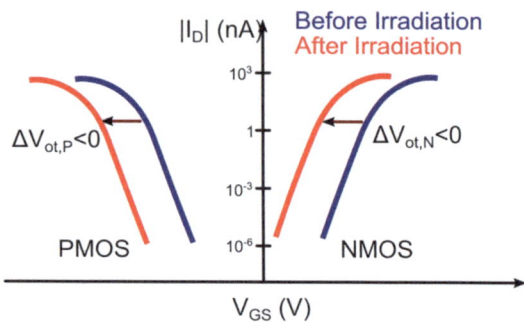

(within picoseconds) when an electric field is applied [45]. During the transport of drift and diffusion, electron-hole recombination takes place simultaneously.

The left holes can move towards gate/SiO$_2$ by applying a negative gate bias voltage or towards Si/SiO$_2$ by applying a positive gate voltage. These holes destroy the local potential field of the SiO$_2$ lattice as they move through the SiO$_2$. When holes move through SiO$_2$ by polaron hopping, they are captured in oxide traps, which are defects or imperfections in the SiO$_2$ material [45, 46]. The oxide traps always lead to a net positive charge in both NMOS and PMOS transistors, which always results in a negative threshold voltage shift. Figure 2.9 shows the threshold voltage shift due to the oxide trap in TID effects. For NMOS, it is more difficult to turn it off and it causes more leakage. In contrast, PMOS is harder to turn on and has a lower current drive strength.

The holes that escape recombination will be transported through the oxide to the Si/(SiO$_2$) interface by hopping [25] when a positive electric field is applied. When the holes arrive at the Si/(SiO$_2$) interface, part of them is captured by the interface trap. Interface traps are created by the defects of the tetrahedral lattice structure of the silicon crystal, which are located at the interface between the semiconductor material and the insulator (Si and SiO$_2$). The interface state can exchange carriers with the silicon substrate, and the exchange rate is closely related to the energy level of the traps, the number of traps and the electrical polarity. In contrast to the oxide traps, the interface traps in p-channel/n-channel transistors are positive/negative [46, 47]. Figure 2.10 describes the threshold voltage shift due to the oxide trap in TID effects.

In addition to the gate oxide, the isolation oxide can also experience electron-hole pair generation and amphoteric interface charge trapping. The two most common types of field-solation oxide in CMOS technology (see Fig. 2.11) are Local Oxidation of Silicon (LOCOS) and Shallow-Trench Isolation (STI) [48]. Similar to the gate oxide, radiation-induced positive charges build up in the LOCOS and STI, forming an n-type region below the field oxide. When the surface inverts, a conductive path is created, resulting in an additional leakage current between the drain and source or between the transistors (see Fig. 2.12). On the other hand, the parasitic depletion region shortens the effective channel length due to the interactions with the source and drain depletion regions, which is referred to as the Radiation-

2.5 Radiation Effects in Semiconductor Devices

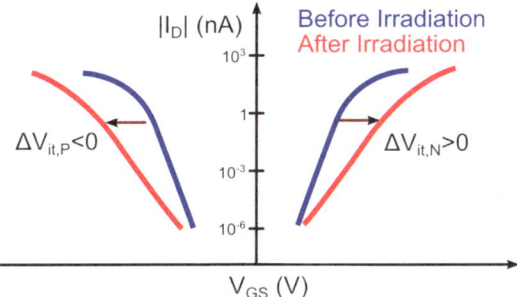

Fig. 2.10 Threshold voltage shift due to interface trapped charge

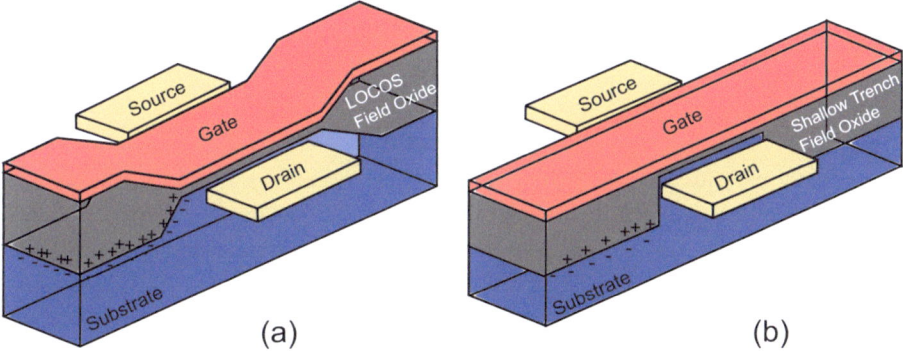

Fig. 2.11 Cross-section of **a** a LOCOS isolated and **b** an STI transistor

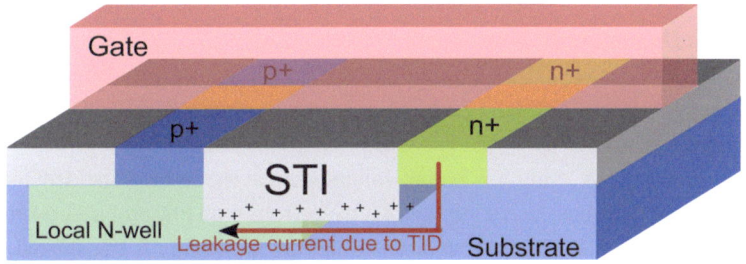

Fig. 2.12 Leakage between two MOSFETs due to TID effects in STI

Induced Short-Channel Effect (RISCE) [49, 50]. Since the radiation-induced charges are positive, the leakage currents mainly occur in NMOS. The source and drain leakage current acts like a parasitic MOSFET connected in parallel with the main transistors.

Figure 2.13 presents the drain-to-source current as a function of the gate-to-source voltage of an NMOS with and without the leakage of the parasitic field oxide transistor due to TID effects. It can be seen that before irradiation, the threshold voltage is positive, and the

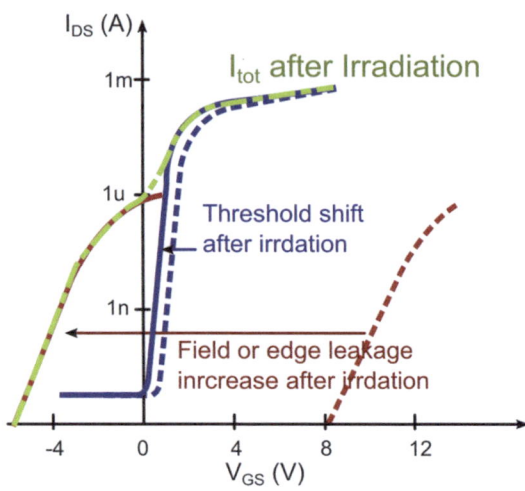

Fig. 2.13 I-V curve of an NMOS transistor, including the current from the gate-oxide and parasitic field-oxide or edge-oxide transistor

leakage of the parasitic field oxide transistor is negligible until V_{GS} is higher than 10 V. After irradiation, the leakage increases dramatically when V_{GS} is less than 0 V due to the parasitic field oxide transistor in STI or LOCOS. This leakage causes static current consumption in the application. In addition, the blue curve (current without parasitic field oxide transistor) shifts to the left, as positive charges are trapped in the gate oxide and the threshold voltage is reduced. Comparing the leakage fraction of the parasitic and the main transistor, the leakage of the parasitic field oxide transistor can increase the leakage by 4 to 5 orders of magnitude, which is the main cause of TID-induced IC failure in advanced technologies [25].

Displacement Damage (DD) is another form of cumulative physical damage caused mostly by neutron or proton dose, which causes electrical/thermal/optical property degradation as the dose increases [12, 23]. Figure 2.14 demonstrates the mechanisms of DD. In semiconductor devices, the silicon substrate is based on single-crystal material growth and has an extremely low defect concentration in both volumes and at the surface. Incident high-energy particles (protons, neutrons, heavy ions and even electrons with low probability) collide with the lattice atoms in the substrate. Part or all of the kinetic energy is transferred to the lattice atoms, causing the silicon nucleus to escape from its correct physical location within the crystal lattice. This introduces a vacancy in the place where the atom was released and produces an interstitial in the place where the dislodged atom settles. The vacancy and interstitial together form a so-called Frenkel pair [51]. Such defects in a crystal result in a local asymmetry in the crystal structure and, specifically for semiconductors, change the electron-hole pair interactions by causing different energy states, often within the bandgap. Normally, most radiation directly leads to ionization. Radiation-induced DD have smaller

2.5 Radiation Effects in Semiconductor Devices

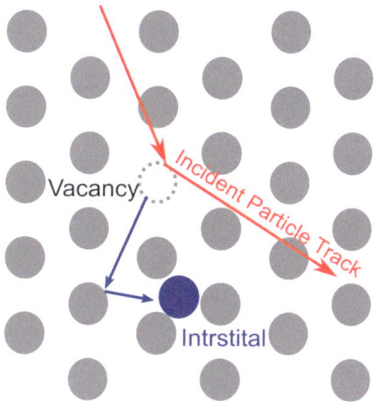

Fig. 2.14 Displacement damage caused by the incident particle

cross-sections[4] compared to ionization since the physical collision must happen between the incident particle and the lattice atoms. Furthermore, a higher kinetic energy is needed to form a vacancy by displacement (approximately 15 eV in silicon) than to create electron-hole pairs (approximately 3.6 eV in silicon) [12].

2.5.2 Single-Event Effects (SEEs)

When a high-energy ion passes through an electronic device and penetrates the semiconductor substrate, the ion leaves a high density of ionized electron-hole pairs in its wake. These additional electron-hole pairs are subject to carrier recombination and/or carrier transport. Normally, these two effects occur simultaneously. The circuits and the physical structure determine the proportionality of the two effects. In charge carrier transport, diffusion and drift are two basic transport mechanisms. The local charge concentration gradient causes diffusion. The charge carriers move from regions of high concentration to regions of low concentration. In comparison, charge carrier drift is driven by the local electric field.

The reverse-biased p-n junction is the most charge-sensitive region in semiconductor devices. Figure 2.15 shows a reversed-biased N+/P diode and the transient current generated by the passage of a high-energy ion. The N+ contact is biased with a positive voltage with respect to the substrate. The depletion zone is formed before ionization. When an energetic ion passes through the depletion zone, a cylindrical track with a high concentration of electron-hole pairs is left in the ion's wake (Fig. 2.15a). Then the electron-hole pairs are immediately separated by the electric field between the N+ contact and the substrate. The negative charges drift towards the N+ contact and produce a large current and voltage

---

[4] Cross-section can be considered in terms of a characteristic interaction area, outside of which the interaction/collision probability drops to zero. A larger cross-section implies a larger area and a larger probability of interaction. Cross-section, typically denoted as σ, is measured in units of area [37].

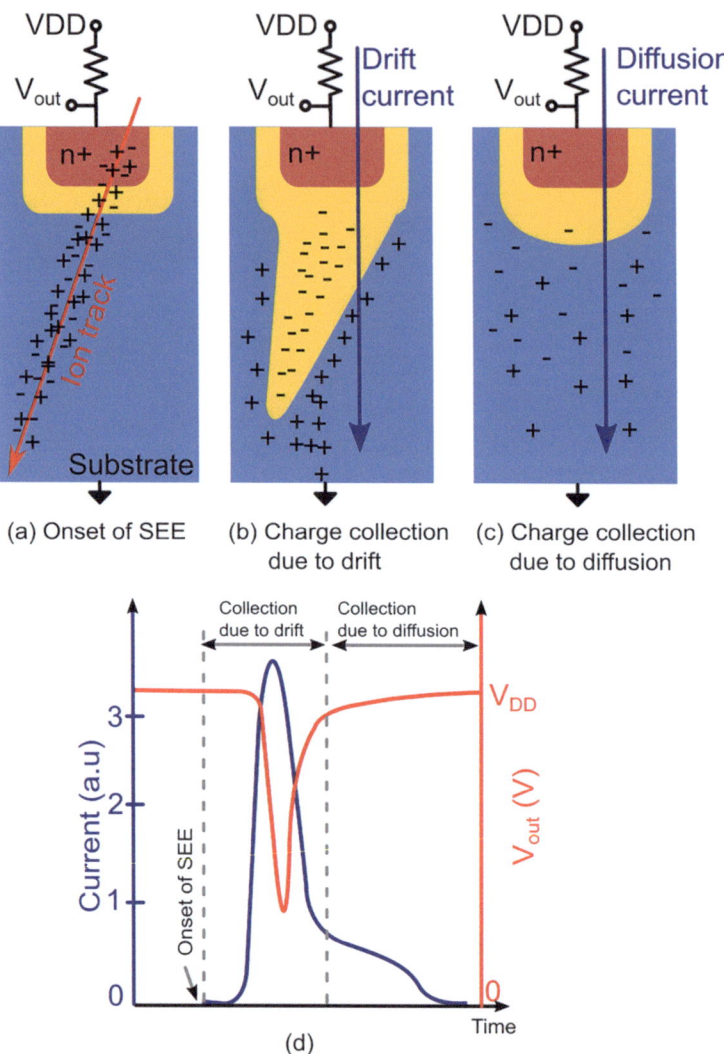

Fig. 2.15 Charge collection in a silicon junction immediately after **a** an ion strike, **b** prompt (drift) collection, **c** diffusion collection, **d** the junction current and voltage induced as a function of time

transition (Fig. 2.15b). After a few nanoseconds of high current from the drift, diffusion begins to dominate the collection process (Fig. 2.15c). Charges in the deep substrate volumes diffuse to the substrate surface with low charge concentration and continuously contribute to the current flow. The complete current curve versus time is shown in Fig. 2.15d and is usually referred to as Single Event Transient (SET) current. The entire SET current profile can last from femtoseconds to microseconds, depending on the design, layout and process. Similar to the N+/P diode, the reversed-biased P+/N diode can also have the same charge

2.5 Radiation Effects in Semiconductor Devices

collection process, but in the opposite direction. However, the N+/P diode usually collects more charge compared to the P+/N diode and has more current if the dimensions of the diodes are the same.

Depending on the location, functionality, circuits, layout, LET and angle of incidence, different SEEs can ultimately be derived. Table 2.2 lists the most important SEEs derived from the archetypal SET event. In general, SETs can be divided into two main categories: analog SETs (ASETs) and digital SETs (DSETs), depending on the type of circuit. The derived subsequent effects can also be divided into soft errors and hard errors. Soft errors are usually random and non-deterministic, which means that they do not permanently damage the affected component. Temporary data corruption or functional errors are the most common soft errors caused by SEEs. In contrast, errors and even permanent damage caused by physical defects or failures are referred to as hard errors.

DSETs occur in combinational logic, clock distribution trees, flip-flops and digital memory circuits. A DSET forms a narrow glitch that can propagate through different stages, and this glitch can be attenuated or widened by each stage [52]. If a DSET is generated in the memory circuits, such as in a Static Random Access Memory (SRAM) or a flip-flop, the memory data may flip due to the DSET. This type of memory data flipping is called Single Event Upset (SEU). If a single memory cell is flipped, this is referred to as a Single Bit Upset (SBU). However, due to the scaling of CMOS technology, the dimensions of a memory cell are greatly reduced and the doping of the substrate is increased. Consequently, Multiple Bit Upset (MBU) in neighboring memory bits increase due to DSET charge sharing in advanced technology nodes. SEUs are a type of data corruption, but the memory circuit itself is not damaged. The SEU is therefore a soft error and can be recovered by rewriting the affected memory cells or avoided by adding redundancy. If an SEU or MBU occurs in an important system register, such as in the control or program execution of Field Programmable Gate Arrays (FPGAs), then the entire system operation can be disrupted and cause a Single-Event Functional Interrupt (SEFI). SEFI is a severe consequence of a bit upset and is also a type of soft (non-destructive) SEE. In addition to SEU caused by DSET, ASET can also cause SEFI in analog circuits. For example, ASET can change the sampled voltage in a sample-and-hold (S&H) circuit . If this S&H circuit is used for the auto-zeroing of an amplifier, the voltage corruption can lead to a complete malfunction of the amplifier until the next auto-zeroing cycle.

Latchup is a well-known and catastrophic reliability problem of CMOS circuits and applications. Parasitic bipolar junction transistors in well-isolated CMOS technology are the origin of latchup [53]. These parasitic BJTs and resistors are illustrated in Fig. 2.16. Under normal circumstances, the collector of the PNP and NPN transistors are connected to VDD and VSS respectively. When a high-energy particle travels through the N-well, electron-hole pairs are generated in its wake. The electrons and holes are then split and drift towards the N+ and P+ due to the electric field. Since the resistance of the N-well and P-well (or substrate) is not zero, the injected current creates voltage drops and causes forward biasing voltage to the parasitic BJTs. In addition, the parasitic BJTs also form a

Table 2.2 Main SEEs derived from the archetypal SET event

Archetypal event	Sub-Events		Digital	Analog	Soft error	Hard error
Single-event transient (SET)	Single-event upset(SEU)	Single-bit upset (SBU)	✓		✓	
		Multiple-bit upset (MBU)	✓		✓	
	Single-event functional interrupt(SEFI)		✓	✓	✓	
	Single-event latchup(SEL)			✓	✓	✓
	Single-event gate rupture/Burnout(SEGR/SEB)		✓	✓		✓

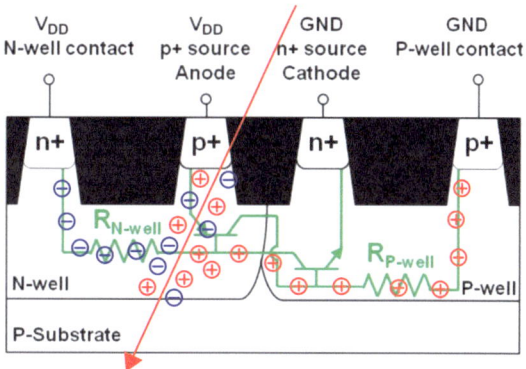

Fig. 2.16 Parasitic BJTs and resistors in a typical twin-well CMOS technology

positive feedback loop. As a result, a low impedance path is formed between the VDD and VSS, resulting in high currents. The SEL can only be eliminated by powering down and shutting off the parasitic BJTs. If the high latchup current continues without disconnection, then physical damage (hard error) will happen to the circuits and system. SEL protection can be an effective method to prevent this damage by monitoring the current from the power supply.

Single Event Burnout (SEB) refers to the activation of the parasitic bipolar structure within a power transistor, usually an n-channel transistor. This activation is accompanied by regenerative feedback, avalanche effects and a high current state. SEB can be highly damaging if suitable protective measures are not in place. SEB occurrences are relatively rare with standard voltage ASICs, except if they are lateral or vertical power devices. Single Event Gate Rupture (SEGR) may happen to power devices but with different mechanisms. It is the destructive fracture of a gate oxide or dielectric layer caused by the impact of a single ion strike. SEGR can lead to a high leakage current in power MOSFETs. As both SEB and SEGR are primarily related to power electronics, they are not the primary focus of this research and therefore no further investigations will be conducted in these areas.

2.5.3 High Dose Rate (HDR) Effects

High Dose Rate (HDR) occurs in special radiation environments where a high dose of gamma rays and/or X-rays is delivered in a very short time interval. The primary effect of HDR events in microelectronics is the generation of global ionization, which subsequently leads to the induction of transient currents in junctions. The microelectronic devices can be catastrophically destroyed by the combination of SEU, SEL, SEB or SEGR [54].

2.6 Radiation Effects Trend in Advanced Field Effect Transistor Technologies

Many variables influence the radiation sensitivity of microelectronic circuits and systems. One of the most important factors is the Field Effect Transistor (FET) technology or process node. As the technology scales, some technologies are more tolerant to certain types of radiation while others are not. Before we look at the radiation effects of the different FET technologies, it is important to give a brief introduction to the main FET structures.

2.6.1 Advanced Field Effect Transistor Technologies

Decades of improvements in the manufacturing process have made Bulk-CMOS technologies smaller and smaller. Unfortunately, leakage currents and Short Channel Effects (SCEs) are the main obstacles to scaling Bulk-CMOS technologies below 22 nm. SOI and three-dimensional channel structures, such as FinFET and Gate-All-Around Field-Effect Transistor (GAAFET), are raised to continue Moore's law, as shown in Fig. 2.17. Several technology structures can be derived from the combination of the two innovations: Bulk-CMOS, SOI-CMOS, Bulk-FinFET and SOI-FinFET.

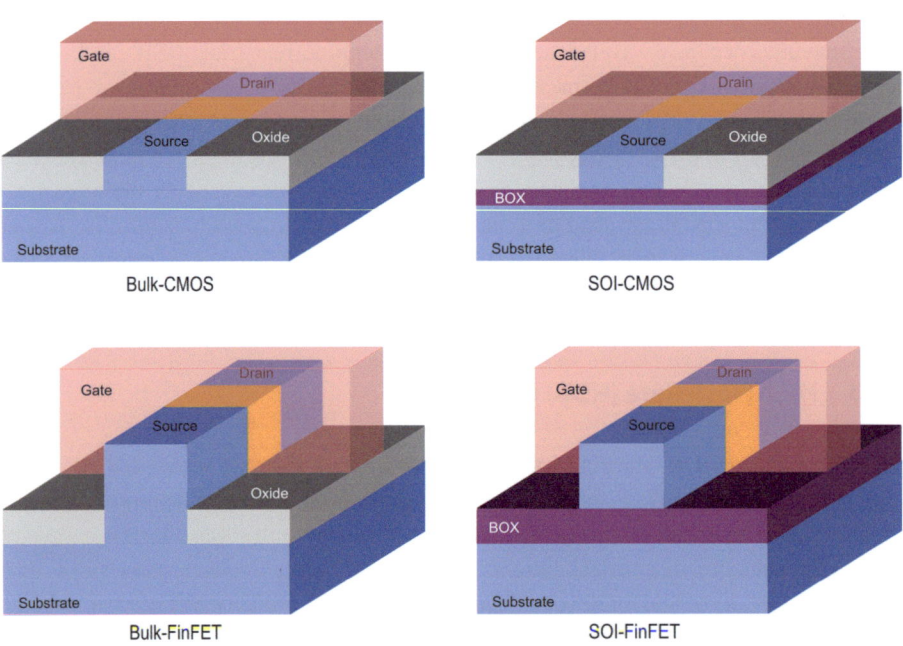

Fig. 2.17 Structure of Bulk-CMOS, SOI-CMOS, Bulk-FinFET and SOI-FinFET transistors

2.6 Radiation Effects Trend in Advanced Field Effect Transistor Technologies

In Bulk-CMOS technologies, the transistors are manufactured on a flat, two-dimensional surface. The current path of the transistor is formed by a gate electrode on a thin insulating layer, which is made of high-k material such as silicon dioxide (SiO_2). The gate electrode controls the current flow between the source and drain regions by modulating the conductivity of the channel region.

Instead of building a transistor directly on a silicon substrate, SOI-CMOS transistors are built on a thin oxide layer, the buried oxide (BOX) (purple region in Fig. 2.17). SOI-CMOS transistors offer advantages in reducing leakage current, improving speed and lowering power consumption. The BOX divides the silicon substrate into top- and bottom-silicon layers. Therefore, an SOI-CMOS transistor inherently consists of a top-gate transistor and a parasitic back-gate transistor. For the back-gate transistor, the substrate acts as a gate contact. Depending on the coverage of the depletion region below the top gate, SOI-CMOS transistors can be divided into partially depleted SOI (PD-SOI) transistors and fully depleted SOI (FD-SOI) transistors. If the depletion region below the top gate extends completely through the top-silicon layer and reaches the BOX layer, it is an FD-SOI transistor. In contrast, PD-SOI transistors contain a non-depleted silicon region between the top gate and the BOX. Therefore, PD-SOI transistors behave to a certain extent like Bulk-CMOS.

When the length of the channel of a Bulk-CMOS transistor is close to the length of the depletion layer in the source and drain junctions, this Bulk-CMOS transistor suffers from SCEs [55]. SCEs have a significant influence on the drain voltage on the channel formation and reduce the threshold voltage. To solve this problem, instead of further scaling the two-dimensional Bulk-CMOS structure, the three-dimensional Fin gate structure was developed, which controls the channel in both horizontal and vertical directions (see Fig. 2.17). By 2002, the Bulk-FinFETs reached the gate length of 10 nm. Gate-All-Around (GAAFET), a new evolution of the FinFET technology, was first introduced in 2006 and scales the gate length further below 5 nm [56].

Although the use of advanced technologies is limited in most space applications due to factors such as cost and complexity, they are becoming increasingly popular due to their low power consumption, high speed and resistance to certain radiation effects. Information and knowledge about radiation effects in advanced technologies beyond Bulk-CMOS is sparse, yet some trends have emerged from the limited data. The next section will mainly focus on the analysis of TID effects and SEEs in different process nodes and transistor structures. The conclusion will be a guide for Radiation Hardening by Design (RHBD) later in this chapter.

2.6.2 TID Effects Trend in Advanced Field Effect Transistor Technologies

As described in the previous section, the trapped charge in the insulators and the oxide-channel interface is responsible for the threshold shift and TID-induced leakage. Therefore,

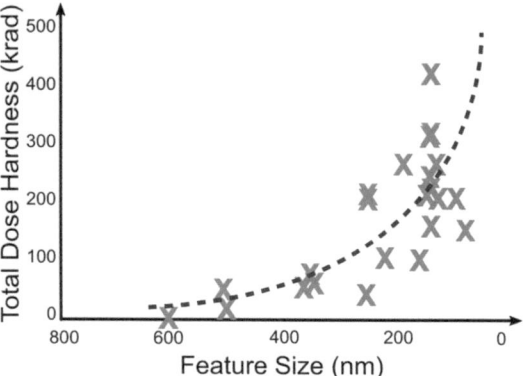

Fig. 2.18 Total ionizing dose failure (functional and/or parametric) level scaling trend for digital CMOS technologies (*Source* Fleetwood [58])

the oxide gate thickness, gate width and length, and supply voltages are the main factors affecting TID sensitivity.

Generally, as the feature size of Bulk-CMOS transistors shrinks, thinner gate oxide and lower supply voltage will cause less trapped charge in the gate oxide and the oxide-channel interface and thus less threshold voltage shift [57, 58]. Compared to the threshold voltage shift, field oxide leakage dominates the degradation and failure due to TID effects because the field isolation oxide is much thicker than the gate oxide. Beyond 350 nm CMOS technology, the LOCOS process has been replaced by the STI process, in which a trench is etched and filled with deposited films to isolate the transistors [59]. STI has better quality and morphology of the deposited dielectrics. Therefore, transistors with STI have a lower edge leakage current due to the parasitic field oxide transistors compared to technologies with LOCOS. Figure 2.18 shows the scaling trend for the TID hardness [58]. As the feature size reduces, both the transistor and functional level tolerances for TID effects increase significantly. As process nodes dropped to 180 nm or below, most circuits could pass 100 krad(Si) even at HDRs. Deep submicron technology nodes (90 nm and below) can basically tolerate 300 krad(Si) or even into the Mrad range [60].

The TID response of SOI-CMOS transistors can be more complex than that of Bulk-CMOS transistors due to the BOX (or back-gate). Similar to the charges trapping in the top gate oxide and the Si/SiO$_2$ interface, charges are also trapped in the BOX layer and the BOX/silicon interface, contributing to a degradation of the transistor characterization. It has been reported that the TID response strongly depends on the charge trapped in the BOX and its relative weight is determined by the coupling effects between the top-gate and back-gate [61–63]. Thus, the bias conditions of the back-gate have a significant influence on the TID threshold shifts. Figure 2.19 shows the threshold voltage shift at different bias conditions: 0V (all terminals are grounded), OFF (drain is connected to VDD and the rest to GND), ON (gate is connected to VDD and the rest to GND) and TG (source and drain are connected to VDD and the rest to GND). The polarity of the threshold voltage shift can even be changed by the bias conditions. By thinning the BOX layer, this bias dependence and the

2.6 Radiation Effects Trend in Advanced Field Effect Transistor Technologies

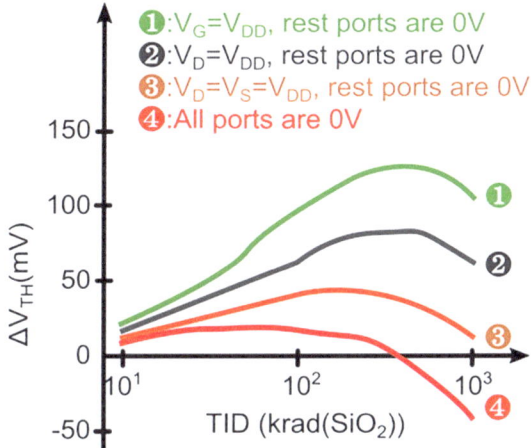

Fig. 2.19 Top gate threshold voltage shift at different bias conditions in an N-type SOI-CMOS transistor with a gate length of 30 nm and BOX thickness of 145 nm

absolute threshold shift are relieved owing to a drastic reduction of the radiation-induced trapped charges in the BOX [63]. In addition to the threshold shift, the leakage current due to TID effects is also complicated. There are two leakage current paths between the source and drain as the trapped charge forms two inversion layers: one below the top gate and one above the BOX layer [64]. The sidewall leakage current path near the STI also contributes to the leakage current between the source and drain, although the inter-transistor path does not exist in FD-SOI structure [65].

For FinFETs, the TID responses follow the same trend from that of Bulk-CMOS transistors. When a FinFET has a diode bias ($|V_{GS}| = |V_{DS}| = $ VDD), the worst TID degradation occurs [66, 67]. The trapped charge in the STI still dominates the TID effects [67]. Figure 2.20 compares the off-state current before and after TID in different commercial FET technologies [68]. In Bulk-CMOS technologies from 150 nm to 45 nm, a continuous decrease in the off-state current (or leakage current) can be found. Thanks to the thinner gate oxide and STI, the leakage current also scales with the scaling of the technology feature size. It should be noted that PD-SOI transistors have similar behavior to Bulk-CMOS transistors due to the lower coupling between the top gate and BOX and the higher doping in the channel compared to FD-SOI transistors. Therefore, PD-SOI transistors have less leakage compared to FD-SOI transistors. However, scaling from 32 nm to 14 nm leads to an increase in TID sensitivity due to the non-planar structure. 14 nm Bulk- and SOI-FinFET and SOI Ultra-Thin-Body and BOX (UTBB) show this reversal in the STI response [68]. As mentioned above, the STI in all MOSFETs contributes to a noticeable leakage path, and it is also the most important factor for the radiation-induced FinFET leakage. In the UTBB structure, the light doping in the channel and the strong coupling between the top gate and the BOX cause this structure to exhibit a huge leakage current after 1 Mrad compared to other 14-nm processes. To summarize, technology scaling of Bulk-CMOS transistors contributes to the

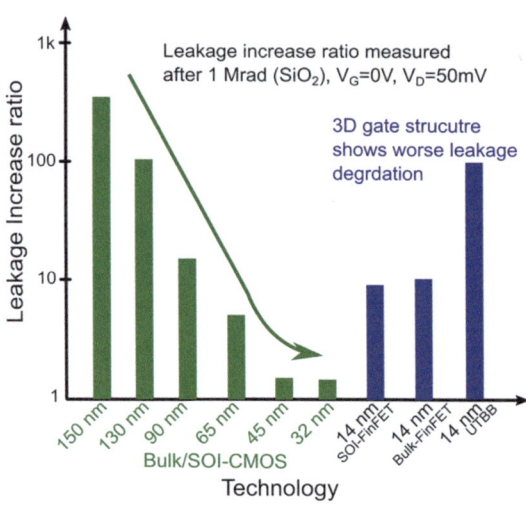

Fig. 2.20 Commercial Technology TID response of off-state current for a dose of 1 Mrad(SiO$_2$) versus technology scaling

reduction of TID sensitivity. However, further scaling with SOI and three-dimensional gate structures leads to lower TID tolerance.

2.6.3 SEEs Trend in Advanced Field Effect Transistor Technologies

As already mentioned, SEEs arise from the electron-hole pairs generated in the wake of the energetic particle. Among all SEEs, SEU and SET are the two main issues in integrated circuits. Therefore, the trend analysis is mainly based on these two effects.

An SEU occurs in digital memory circuits such as SRAM and DRAM. To evaluate the SEU trend in different technologies, the SEU response of an SRAM is placed at the center to simplify the analysis. The critical charge (Q_C), threshold LET (L_T) and saturated cross-section (σ_∞) are commonly used to characterize the SEU sensitivity of SRAM circuits. Q_C is the minimum amount of charge and L_T is the minimum LET required to generate an SEU event. σ_∞ (with the unit of cm^2/b) refers to the maximum cross-section for a single ion strike or particle impact that causes an SEU in an SRAM cell with arbitrarily high LET. Figure 2.21 shows an example of the measured cross-section compared to the LET in a 0.5 /mum PD-SOI SRAM. The L_T and σ_∞ can be determined from the observed data by Weibull fitting.

The SEU responses (Q_C, L_T and σ_∞) as a function of different technologies (l, channel length) are shown in Fig. 2.22 [70]. Figure 2.22a shows that the Q_C decreases continuously with l. The slope for Bulk-CMOS transistors is proportional to $l^{1.5}$, and for SOI-CMOS transistors it decreases with a slope of l^2. The $l^{1.5}$ slope for Bulk-CMOS transistors is caused by two factors. First, the gate capacitance C_g has the same scaling ratio l^1 as the technology. Secondly, the supply voltage is roughly scaled with a ratio of $l^{0.5}$. The combination of these

2.7 Radiation Hardening Techniques

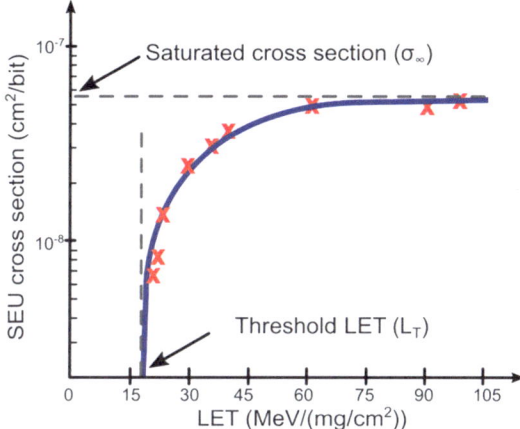

Fig. 2.21 The SEU cross-section as a function of LET and the Weibull fitting for 0.5 μm PD SOI SRAM (Data from Liu et al. [69])

two factors leads to $l^{1.5}$. For SOI transistors, Fig. 2.22a shows that they have less Q_C than Bulk-CMOS transistors with the same l. Although the BOX in the SOI structure can reduce the collected charge at the drain node of the transistor, SOI-CMOS transistors have a lower parasitic (junction) capacitance. Consequently, the SOI structure is more sensitive to the ionization charge from single-particle SEE compared to Bulk-CMOS transistors. As for the FinFET gate structure, its charge tolerance becomes better. It has a higher Q_C compared to the planar structure (Bulk-CMOS and SOI-CMOS structure) because the Fin structure has a relatively large parasitic capacitance due to its inherent three-dimensional gate structure. This parasitic capacitance is undesirable in technology scaling as it reduces the speed of operation. But it increases the robustness to SEEs.

As for the SEU threshold LET (L_T), it can be seen from Fig. 2.22b that the SOI- and Bulk-CMOS technologies have almost a similar decreasing slope ($l^{1.5}$). Recent literature reports that SOI-CMOS technologies largely deviate from this trend and can exhibit up to a 150x increase. But in general, SOI-CMOS transistors have a higher L_T compared to Bulk-CMOS transistors because Bulk-CMOS transistors can collect the charge from the deep substrate, while the BOX layer in SOI-CMOS transistors isolates the deep substrate.

Figure 2.22c shows the saturated cross-section σ_∞ of SRAMs against the transistor length. The slope of the trend is l^2, which is also the ratio of the decrease in unit cell area. Note that two lines are drawn in the graph: the unit cell area of SRAM A_{cell} and the typical area of an ion track A_{ion}. Normally A_{cell} should be the upper limit of σ_∞. However, some data for Bulk-CMOSs exceed this limit and give the MBUs. This happens mainly when the transistor length is shorter than 100 nm. In contrast to Bulk-CMOS transistors, SOI-CMOS transistors persist and show more distance to the A_{cell} boundary. This is due to the help of BOX layer isolation. There is a report that the Bulk-FinFETs have reached the A_{ion}. This indicates that σ_∞ will finally stop decreasing and converge towards A_{ion} in the future.

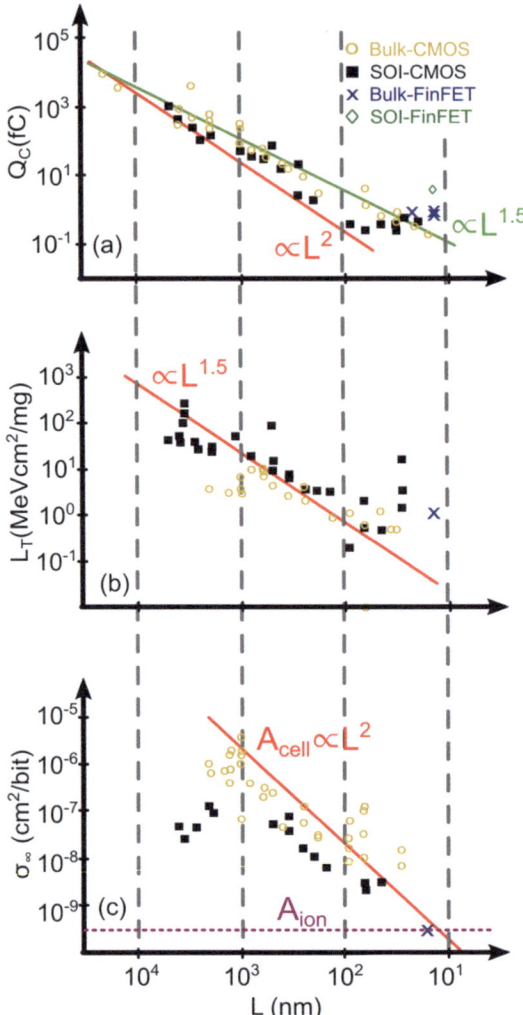

Fig. 2.22 SEU response of the SRAM cell in different technologies: **a** critical charge **b** threshold LET **c** saturated cross-section of the SRAM versus the channel length [70]

To summarize, the scaling of technology has made transistors more tolerant to TID effects due to thinner gate oxide and advanced STI. However, the shorter channel length makes them more sensitive to SEEs as the power supply and parasitic capacitance are reduced.

2.7 Radiation Hardening Techniques

Proper shielding can be an effective method to mitigate or even eliminate the radiation effects of microelectronic devices. However, in space applications, shielding requires more mass and space for the launch payload and incurs additional launch costs and risks. Therefore,

2.7 Radiation Hardening Techniques

chip-level radiation hardening is preferred for space projects. As illustrated in Figure 1.3, the tolerance of microelectronic chips and applications such as ADC performance can be improved from four aspects: technology, circuits, architecture and system. The four aspects can be divided into two categories: technology hardening is Radiation Hardening by Process (RHBP), and the other three aspects (hardening in circuits, architecture and system) belong to RHBD.

2.7.1 Radiation Hardening by Process (RHBP)

Radiation sensitivity can be reduced by changing the base semiconductor process. One example is replacing the base substrate with a material with higher conductivity [71]. The highly doped substrate significantly reduces the substrate resistance and also reduces the SEL sensitivity. However, since the electrical model of the base process is the most accurate, any process change can alter the electrical properties of the devices and affect the design performance. To minimize the electrical drift caused by the process modification, an epitaxial (epi) layer with the base doping is applied over the modified substrate (with the same transistor dimension). The thicker the epi-layer, the lower the electrical drift, but the less SEL hardening effectiveness (Fig. 2.23).

However, with the scaling of the technology and the complex 3D structure for MOSFETs, the process flexibility becomes more and more limited. Therefore, modifying the process is not a first option. Instead, one can choose a suitable technology instead of modifying it. For example, the SOI process is more tolerant to SEEs compared to Bulk technology. So, adapting the existing design from Bulk technology to SOI technology to increase SEE tolerance does not cost any process change or electrical parameter drift.

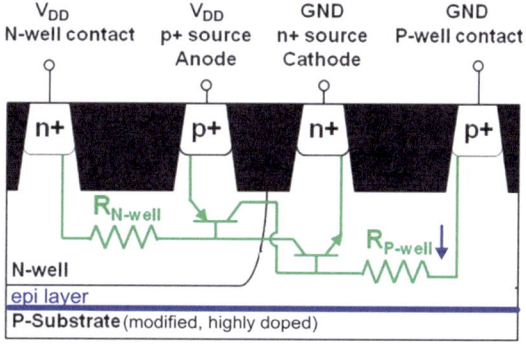

Fig. 2.23 P-type epi layer on top of the highly doped substrate to increase the SEL tolerance

2.7.2 Radiation Hardening by Design (RHBD)

Compared to RHBP, RHBD takes advantage of physical and dimensional properties to improve tolerance to radiation effects. RHBD has no impact on the process and places no special demands on foundries. Since most space-related chips are produced in small quantities, commercial processes without customization together with RHBD are the most cost-effective solution for space chips. RHBD can be applied at different design levels: transistor, architecture and system. But all have disadvantages in area, power or speed.

2.7.2.1 Layout Level

One of the simplest solutions is to increase the transistor dimensions in order to improve the SEE tolerance. A larger width increases the driving strength of the transistor, which lowers the resistance to ground or supply. When the energy particle creates the electron-hole pairs in the transistor, the transistor with higher drive strength has a smaller SET magnitude and duration [72]. Torrens et al. [73] has used cells with six-transistor SRAM cells with upsizing PMOS and downsizing NMOS to minimize the SEUs by a ratio of two. However, when this approach is applied to large amounts of digital or memory circuits, the increase in dimensions results in a larger area and higher power consumption during switching.

Another method is to use specific layout strategies when dealing with different radiation effects. Placing a guard ring (illustrated in Fig. 2.24) between NMOS and PMOS with additional well contacts has been proven to be an effective method for attenuating SEL [74, 75]. The guard ring can also reduce the charge sharing between the transistors due to SEE and reduce the SET and MBU probability [76]. In addition, placing a deep N-well (DNW) can also be an effective method to reduce charge collection within the transistor and charge sharing between transistors. DNW has been proven to have an average 2x reduction of charge collection in 65 nm core translators in [72]. Nevertheless, adding the guard ring or DNW increases the total area of the layout area and increases the distance between the circuit nodes. This not only affects the chip area but also the operating speed.

When hardening the chip against TID effects, Enclosed Layout Transistors (ELTs), shown in Fig. 2.25, can be an effective solution when the technology is greater than or equal to 130 nm. As Bulk-CMOS scales further, the main TID leakage is caused by the transistor edge. The ELT structure interrupts the leakage path between adjacent n+ junctions with different potentials [77, 78]. Nevertheless, the ELT structure is not free of charge. In addition to the area penalty, the ELT structure poses another challenge in creating an accurate electrical model due to the asymmetry. To obtain an accurate model, more parameters need to be considered. As the technology scales further below 90 nm, the ELT structure is rarely used because the edge effects are largely scaled.

2.7 Radiation Hardening Techniques

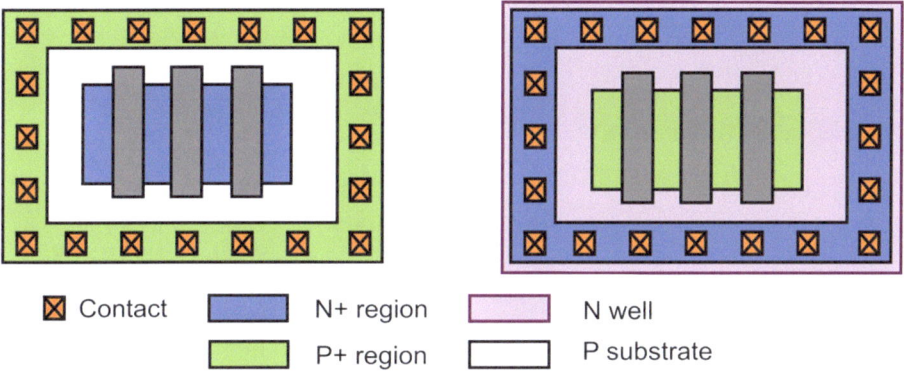

Fig. 2.24 P+ and N+ guard ring for NMOS transistor and PMOS transistor

Fig. 2.25 ELT structure and its model parameters

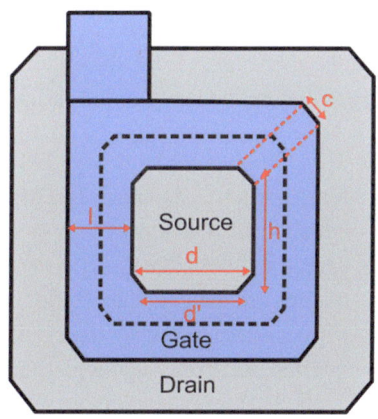

2.7.2.2 Circuit Level

Above the circuit level, the RHBD can be applied to the circuit for a function block. By adding additional transistors, feedback or even reforming a novel structure of the functional block, the radiation hardness can be improved. The Dual Interlocked Storage Cell (DICE), shown in Fig. 2.26, is a successful example of improving the storage SEU tolerance. The DICE structure was first introduced in [79] with a 12-transistor implementation. Compared to the conventional 6-transistor latch, the DICE structure has two feedback paths. When an ion impacts one of the four sensitive nodes, the other transistor provides compensation and retains the data. However, this hardening is valid only when only one node is affected. A sufficient distance must, therefore, be maintained between the transistors in the layout. The DICE structure requires twice the area and power, but reduces the SEU rate by more than 1000 times [80].

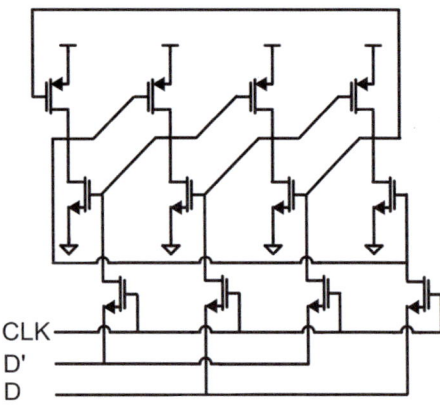

Fig. 2.26 Schematic of The DICE memory cell

2.7.2.3 System Level

At the even higher level, the system level, RHBD can also be applied. This can be achieved by voting, redundancy, cold sparing, algorithm, etc. [81].

- Triple Module Redundancy (TMR) shown in Fig. 2.27a is a widely used voting technique for radiation hardening [82]. TMR can be used not only at the system level, but also at the circuit level. The same block or system is duplicated three times and the same process is carried out simultaneously or with a slight delay (redundancy in the time domain). The outputs are then compared and, in the event of a discrepancy, the system can detect and correct errors, ensuring high reliability and resistance to hardware failures and transient interference from radiation. TMR is mostly used in the digital domain, but can also be applied to the analog domain if the output selector is replaced by an averaging circuit. The error amplitude is then divided by a factor of three and wins an SNR of 9.5 dB. Similar to other hardening techniques, TMR is not free and triplicating means tripling the power, area and cost.

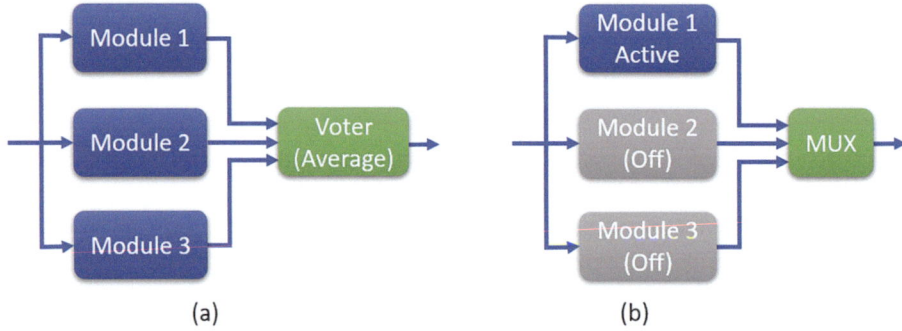

Fig. 2.27 System level hardening techniques: **a** TMR technique **b** spare mode technique

- Spare mode techniques in Fig. 2.27 bat the system level also use several identical blocks, but not simultaneously. Only one of them is active and executes the operation. The other blocks remain in cold spare mode, i.e. switched off. If the malfunction is detected in the active block, the swap from the compromised block to the replacement block is carried out. Since the change from the switched-off state to the active state can take some time, fast swapping can be achieved by the hot spare, which consists of a spare block in the switched-on state without input, while the other spare blocks are switched off. The spare mode techniques can be used to extend the life of the mission when it is limited by TID effects, as the electric field has a significant influence on the TID effects. Switching off can slow down the degradation caused by TID compared to the active state [62, 66, 68]. If the performance of the activated block is no longer acceptable, the mission can be continued within the specification requirements by switching to the replacement block. In addition, spare mode techniques can also be used against SEEs, such as SEU, SEFI and SEB.
- Integrating error detection and correction (EDAC) algorithms into a digital block or memory circuit can also reduce the data error rate and malfunction probability. These algorithms include adding additional sign bits, converting the data to a specific code pattern, etc. [83] encodes the data in SEC-DED-DAEC code to tolerate single and double non-adjacent errors. Normally, the EDAC algorithm requires more memory locations in the memory circuits compared to the original codes, and additional circuits are needed for encoding and decoding. Therefore, more area and power is needed for hardening the algorithm.

2.8 Conclusion

In this chapter, the physical mechanisms that cause radiation effects were briefly introduced. Then the two main effects on semiconductor devices—TID effects and SEEs—are examined. The review of radiation effects in different technology nodes revealed that the transistors in scaled technologies are more tolerant to TID effects but more vulnerable to SEEs. Therefore, in recent years, radiation hardening has focused more on SEEs in technologies below or equal to 90 nm. The following chapter and design also put more emphasis on SEEs. Last but not least, techniques for attenuating SEEs at different design levels are presented. Some of the techniques are applied to the radiation-hardened ADC design in the later chapters.

Radiation Hardened IC Design and Evaluation Flow 3

Abstract

In this chapter, we present the process of performance evaluation of CMOS technology against radiation effects, which serves as a fundamental cornerstone for the subsequent radiation-tolerant design. We then present a comprehensive radiation-tolerant ASIC/IC design flow that builds on the conventional ASIC/IC design process and includes additional steps to account for radiation effects. All tests and steps ensure both electrical performance and radiation tolerance.

3.1 Introduction

The first part of this chapter introduces the process of technology assessment to evaluate the response of a technology to radiation effects. In the conventional ASIC and IC design flow, the electrical information of the technology is provided and integrated into the Process Design Kits[1] (PDKs) provided by the foundry. Therefore, an evaluation of the electrical technology is not mandatory. However, the foundry's PDKs usually do not contain any information on radiation characterization. In order to use the process for radiation-related ASIC and IC designs, a detailed understanding of radiation must be established through the evaluation process. In the second part of this chapter, the design flow for radiation-hardened ASICs and ICs is presented. The flow is based on the conventional design flow with additional steps and checkpoints.

---

[1] A PDK is a set of files used in the semiconductor industry to model a manufacturing process for the design tools used to design an integrated circuit.

© The Author(s), under exclusive license to Springer Nature Switzerland AG 2026
Z. Li et al., *Radiation Tolerant Nyquist Analog to Digital Converters*, Synthesis Lectures on Engineering, Science, and Technology,
https://doi.org/10.1007/978-3-031-95599-0_3

3.2 Performance Evaluation Flow for CMOS Technology to Radiation Effects

If a technology has never been used for radiation-hardened IC or ASIC development, the first task is to perform the evaluation of the radiation effects. The assessment aims to understand the response of the technology or process to radiation effects. Although the trend of the technology's response to radiation has been analyzed in the previous section, some subsequent effects may still be beyond expectations. More importantly, quantifying the effects is critical to achieve sufficient hardening and avoid over-hardening the design. Figure 3.1 describes a general flow for the evaluation of a CMOS technology. This evaluation flow can also be applied to other semiconductor technologies, but in this context, we mainly deal with CMOS technology. It can be briefly divided into three categories: exploration chip design, radiation testing and hardening against radiation effects.

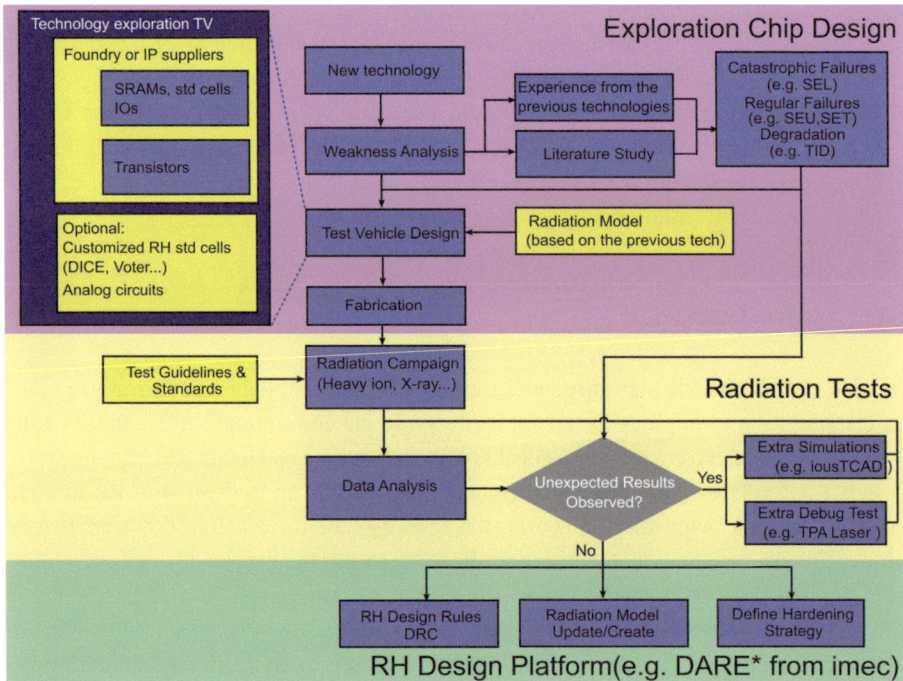

DARE = Design Against Radiation Effects

Fig. 3.1 Evaluation flow for characterization of CMOS technology under radiation

3.2.1 Exploration Chip Design

While previous research may have explored the target process, primarily documented in research papers, electrical performance and radiation behavior can exhibit variations with different foundries and recipes (high voltage, low power, RF, digital, etc.). Therefore, it is common to develop and evaluate a test chip with the target technology to gain an accurate and quantitative understanding of a technology's radiation performance.

When analyzing the performance of the technology, a thorough literature review of the target process can provide a valuable initial overview of potential bottlenecks in radiation tolerance. The evaluation points can be divided into catastrophic failures, regular failures and regular degradation. As explained in the previous section, the most important radiation problems in modern CMOS technology are the SEEs and TID effects. Among these effects, SEL is one of the most fatal problems at the transistor level as it can shorten the path between supply and ground. Consequently, SEL can generate a huge current and burn out the circuit if no countermeasure is taken. The only way to recover from SEL is to turn off the chip, thus interrupting its functionality. Therefore, SEL is treated as a catastrophic failure in the radiation evaluation points. The test case for SEL must be included in the test vehicle to determine its threshold LET value. Similarly, other catastrophic failures can also be defined depending on the technology and radiation type. In addition to catastrophic failures, regular failures and degradation such as SET, SEU, MBU, threshold shift and leakage increase of TID effects must also be considered in the development of the test vehicle. These types of failures and degradation are expected during irradiation. Therefore, knowing their quantitative value in the technology is crucial for future development and hardening.

Once the evaluation points have been determined, the design of the test vehicle can begin. The initial exploratory test chip (or test vehicle) typically implements the existing general-purpose ICs and libraries, which include but are not limited to SRAMs, the standard cell library, and the Input/Output (IO) library. These ICs and libraries usually come from foundries or IC vendors and are used for general design purposes without any radiation hardening. Using existing ICs and circuits is always preferred, as it can reduce design effort and cost. However, their radiation performance must be evaluated. Some ICs or designs may not be sensitive to certain radiation effects and can be used without hardening. Therefore, their SEE and TID tolerance needs to be quantified. In addition, basic transistor victim devices are also interesting to test. Quantifying the electrical response to radiation effects, such as the SET electrical parameters in the time domain and the TID degradation factor, helps to complete the radiation model for the subsequent design. By implementing the transistors in different types, dimensions, structures and layouts, the sensitivity of the transistors can be investigated. Last but not least, some previous silicon-proven radiation-hardened designs can also be included, such as the DICE flip-flop, TMR structure, to check whether they are still effective in the target technology.

When designing an exploration test chip, a radiation model can be very helpful in estimating the performance. Several tools have been developed to model the response of semiconductors to radiation effects. Technology Computer-Aided Design (TCAD) is one of the most commonly used tools for mimicking and extracting numerical details of semiconductor response [84]. However, TCAD simulation is more suitable for testing the behavior of a single transistor than for detailed design verification, as the design may contain multiple transistors that require enormous computing power. Therefore, for complicated analog-mixed signal designs, it is a more efficient method to perform the simulation with the Spectre-based tool [85, 86]. For example, the voltage and current responses of SET were simulated with Analog FTU Hardware Debugging System [85]. The TID-PDK published by CERN allows the developer to select models for TID levels of 100, 200 and 500 Mrad (SiO_2), established for TID experiments carried out at different temperatures [87]. The SET striker in the Design Against Radiation Effects (DARE) Analog Design Kits (ADKs) of the Interuniversity Microelectronics Center (IMEC) is capable of running in the Cadence Virtuoso Analog Design environment and is widely used in radiation-hardened analog/mixed-signal design simulations [88]. In addition, DARE ADKs also provide extra process corners to describe the change of parameters, such as transistor threshold shifts and leakage, due to TID effects [89]. Since the radiation effects in microelectronics strongly depend on the process, the parameters of the radiation model can be roughly derived from the previous technology with a scaling factor. Even if the model is not accurate, it can provide a general estimate of the response of the circuit.

3.2.2 Radiation Testing

After receiving the test chip and producing the test board, radiation tests can be performed. Table 3.1 summarizes the radiation test types, features and test restrictions [90]. Depending on the target application and project, a suitable radiation source can be selected from the table. It should be noted that different radiation sources may have different test restrictions. Therefore, a suitable chip package and Printed Circuit Board (PCB) design must be considered in the design phase, as the requirements vary depending on the radiation sources, equipment and instruments. TID tests often use a Cobalt-60 or X-ray generator to investigate TID-induced degradation. The main limitation of the X-ray source is that the penetration depth of the protons is low and irradiation must be performed at the wafer level or with delidded devices [91]. Therefore, packaging with an open lid or direct bonding on PCB is required. After the X-ray irradiation test, the annealing effect at high or room temperature is used to check the recovery of the test chip. The test process should also consider the flexibility of heating the victim chip and monitoring the temperature with a temperature sensor. A laser test can only be performed from the backside of the chip, as the package and the upper metal layers of the chip would block the laser beam. Therefore, holes are required on the back of the package and PCB. Further requirements from various tests are contained in Table 3.1.

3.2 Performance Evaluation Flow for CMOS Technology to Radiation Effects

Table 3.1 Summary of the radiation tests

Test type	Radiation source & test approach	Main features & limitations
TID	Photons [92, 93]:, gamma rays Co-60 ($\sim$ 1.2–1.3 MeV), Cs-137 ($\sim$ 0.7 MeV);	• High penetration of the source (particularly Co-60) • High capabilities in testing big systems
	X-ray [92, 93]: (keV to several MeV)	• Limited penetration of particles (opening packages required) • High dose rate
SEE	Protons [94, 95] (tens of MeV to GeV))	• High penetration of particles • Limited linear energy transfer (LET) • Component on PCB may be affected (shielding for non-testing components)
	Heavy ions: standard ($<$ 10 MeV/n); high (10–100 MeV/n);	• Limited penetration of particles (opening packages required) • Test in vacuum chamber • High LET, up to 100 MeVcm^2/mg
	Heavy ions [96]: very high energy (0.1–5 GeV/n)	• Moderate penetration • High LET, homogenous LET limited to 30 MeVcm^2/mg.
	Heavy ions [97, 98]: ultrahigh energy (5–150 GeV/n)	• High penetration. • Limited LET: around 10–15 MeVcm^2/mg
	Laser [99] (energy from pJ to tens of nJ)	• The energy is not directly related to LET • An alternative method to debug SEEs • Provide spatial information of the SEE failure • Can not penetrate package and metal layers (die backside must be exposed)
Mixed	CHARM facility [100]: high energy hadrons (up to 24 GeV)	• An alternative approach to mimic the target environment • High penetration • LET is limited to around 15–17 MeVcm^2/mg

The radiation test can produce a large amount of test data. First, the test results must be compared to the original assumptions and predictions, especially for the catastrophic failures that were defined in the design phase of the exploration chip. If the occurrence of catastrophic failures is not consistent with the literature or previous reports, additional debug tests or simulations must be performed to further investigate these effects. In the SEE test campaign, if an unexpectedly high current is observed, the most likely cause is SEL. For example, the authors of [101] observed an unexpectedly increasing rate of SEL for FinFET transistors in the proton beam test. Later, the authors confirmed the increased sensitivity with an additional neutron beam experiment. By building a TCAD model and performing simulations, the authors were able to prove that the increased sensitivity is due to the STI dimension in the FinFET, which is three times larger than the STI in CMOS transistors. This additional debug test and the simulation can be continued until all data can be correctly explained [89].

3.2.3 Hardening Against Radiation Effects

With the conclusions drawn from the test data, the designer can already build up a comprehensive understanding of the target technology. From this, several guidelines and rules for the future radiation-hardened design can be derived. These rules can be written into a script and integrated into the Electronic Design Automation (EDA) tools to automatically check the design. Figure 3.2 shows the schematic design rules used in the radiation-hardened ADK of the DARE platform, which can remind the designer not to use the radiation-sensitive devices and to choose proper dimensions. In addition to the rules and design strategy, the radiation model used in the exploration phase of chip design can be updated with the radiation test data. As explained in the previous subsection, the initial radiation model is built based on the previous technology data with a scaling factor. This initial model may overestimate or underestimate the radiation effect. Therefore, after updating to a more accurate model, this model can be a crucial design tool for the subsequent radiation-hardened design. In the IMEC DARE-180 platform, the ADK models the geometry effects of the ELT transistors and their TID effects. The data of these effects are all extracted from the exploration chip measurement [102].

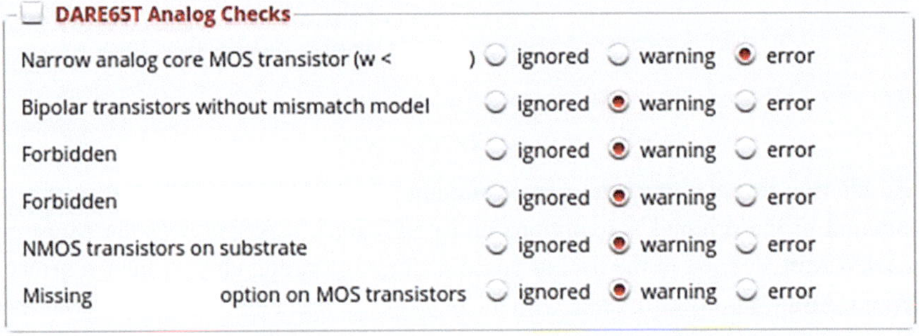

DARE65T Analog Checks	ignored	warning	error
Narrow analog core MOS transistor (w <)	○	○	●
Bipolar transistors without mismatch model	○	●	○
Forbidden	○	●	○
Forbidden	○	●	○
NMOS transistors on substrate	○	●	○
Missing option on MOS transistors	○	●	○

Fig. 3.2 The design rule setup in Cadence Virtuoso to improve the radiation hardness

3.3 Design Flow for Radiation-Tolerant Library, IC and ASIC

There are two main steps for radiation-tolerant IC or ASIC development: design and test. Both processes are based on the conventional flow of IC and ASIC development with additional steps to consider radiation tolerance. Besides, both processes rely on the information and knowledge gained from the results of the radiation effects evaluation results of the respective technology.

3.3.1 Design Flow

The detailed design flow for radiation-tolerant integrated systems is shown in Fig. 3.3. This design flow can also be applied to the radiation-tolerant library and the design of ASIC. When an IC request is received, the first step is to define the specifications of the IC. The specifications can be divided into two parts: electrical specifications and radiation specifications. Normally, both parts are determined by the project and the applications. The electrical specifications contain the working conditions (e.g., voltage supply and temperature range), the performance (e.g., bandwidth, gain and phase margin for an amplifier), and limitations (e.g., power and area limitations). The radiation specifications often include the total dose for TID effects and the LET threshold, the cross-section for SEU or SET. Based on the specifications, the design can be started at the system level. Similar to the conventional design flow, the system structure can be defined based on the specifications. If there is a potential risk of radiation-induced failure at the system level, the IC should be hardened at the system level, e.g., by a hot/cold spare scheme or a more robust algorithm. Sometimes a novel structure has to be invented to fulfill both specifications.

Once the structure has been confirmed by the simulation at the system level, the design can begin at the block level. In each IC, there are several sub-blocks that need to be designed. Since every single block has a different importance for the system, some of them can be very important for the stability of the system. If no RHBD is applied to such a block, it can lead to a catastrophic failure of the entire IC. For example, if a heavy ion penetrates the bias circuits of an amplifier, the SET voltage fluctuation in the bias circuits can cause a huge error in the output, as the unstable bias voltage can easily drive one or more transistors out of the correct operating region. In addition, conventional low-current biasing can make the situation worse, as it can take several or tens of microseconds to return to normal. During this time, the amplifier will not have proper outputs and may cause SEFI to the system, for instance, an ADC. Therefore, the bias circuits of the amplifier need to be hardened. Hardening at the block level can be carried out using the methods described in the previous chapter.

As for schematic design at the transistor level, some transistor types or parameters are warned against or restricted due to the radiation design rules in the EDA tool. Although the flexibility in the selection of transistors for the design is limited, the design at the transistor

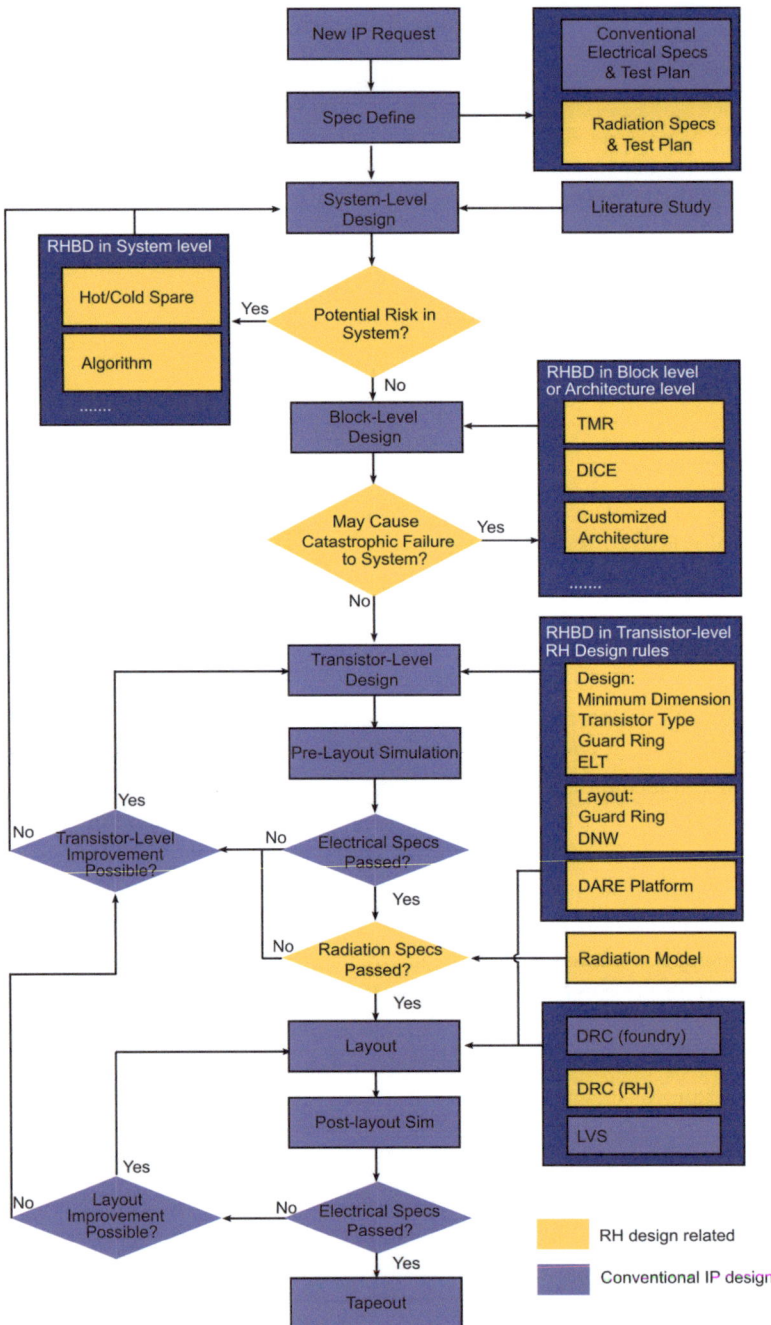

Fig. 3.3 The design flow for radiation-tolerant ICs

3.3 Design Flow for Radiation-Tolerant Library, IC and ASIC

level automatically takes the RHBD into account. For the digital circuits, the compiler can use the radiation-protected digital library when compiling in the back-end flow, such as the DARE library from IMEC, to generate the radiation-hardened digital block directly. In addition to the electrical parameters, the designer can also use the radiation model to mimic the block's response to radiation effects. Since both the radiation and electrical specifications must be met, the transistor-level design may not meet both requirements at the same time. Then the designer must go back to the system level to find an alternative solution. The final step is to draw the layout and perform the post-layout simulation. In the layout, the additional design rule checks (DRC) for radiation hardening can improve the radiation tolerance by including more constraints such as DNW and guard ring in the layout. As with the conventional design steps, the parasitic capacitors and resistors are also included in the design to perform a final check. Once all simulations are passed, the design is ready to be sent to the foundry for production.

3.3.2 Test Flow

The test flow for a radiation-tolerant IC is shown in Fig. 3.4 and is similar to the radiation test phase for the exploration chip. The first step is to perform the electrical test for the IC to check the functionality and electrical performance. If the key function or performance

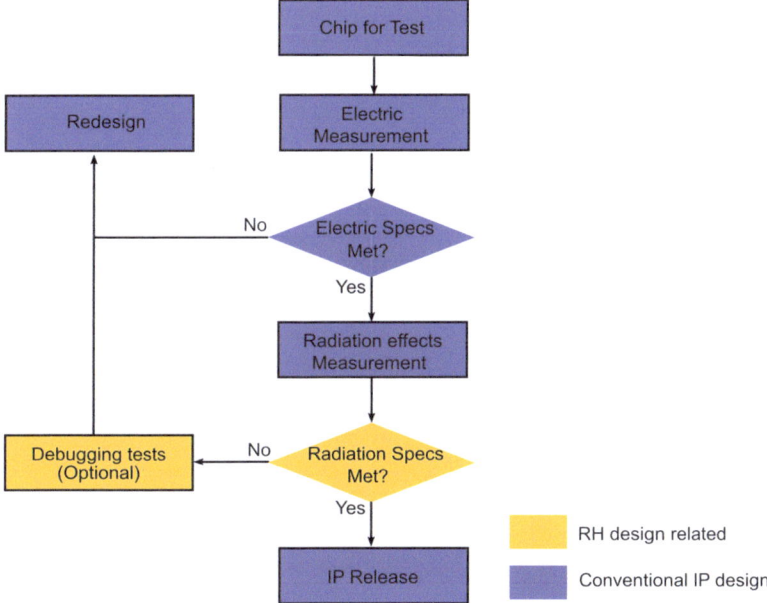

Fig. 3.4 The test flow for radiation-tolerant ICs

does not meet the design specification, a redesign can be initiated to correct the design flaw. Whether the radiation test is continued is not entirely dependent on the results of the electrical test. If only part of the chip is working, the radiation tests can still be performed. However, the decision also depends on the cost, time and manpower of the project. If the radiation test campaign of the target IC shows unexpected results, additional radiation tests can be performed for further troubleshooting, such as a laser test. Once all test results have been clarified, the redesign can begin. Depending on the problem, the design may be rolled back to any design level (system level, block level or transistor level). New radiation restrictions may also be found during the test campaign. In such a case, additional design rules or strategies can be added to the EDA tools.

3.4 Conclusion

This chapter first presents the performance evaluation flow for CMOS technology against radiation effects. This evaluation serves as a fundamental cornerstone for the subsequent radiation-hardened design. Following this, we present a comprehensive radiation-hardened ASIC/IC design workflow that builds on the conventional ASIC/IC design process by incorporating additional steps to account for radiation effects. All checks and steps ensure both electrical performance and radiation tolerance.

Technology Evaulation Design Consideration and a 65-nm CMOS Technology Test Vehicle (Godzilla) for SET Evaluation

Abstract

Since a 65 nm CMOS technology is used for the development of the radiation-hardened ADC, the technology performance under radiation effects must be evaluated before it is used, especially for SEEs. Therefore, this chapter first discusses the advantages and disadvantages of advanced CMOS technology in terms of radiation effects, cost, and development effort. Then, 65 nm CMOS technology is analyzed and evaluated as a suitable candidate for space project development. A 65 nm test chip to characterize the SET ionization charge and pulse duration was developed and tested under a heavy ion beam. The results of the heavy ion test are very valuable to obtain an accurate SET model for radiation-hardened IC, which is later used in the Radiation Hardened By Design (RHBD) process in ADC.

4.1 Introduction

CMOS technology has advanced toward higher integration density, speed, and energy efficiency, but this progress has also increased process complexity and device sensitivity to variations and radiation. Understanding and accurately modeling radiation-induced performance degradation are therefore essential for efficient radiation-hardened design, making precise measurement a critical step. This chapter discusses the general methodology and key considerations for technology evaluation and presents a 65 nm test chip as an example.

4.2 The Methodology of Technology Evaluation Under Radiation

The previous chapter has discussed the performance evaluation flow for CMOS technology to radiation effects, which uses a test chip, or test vehicle, to check the radiation degradation.

Generally, the technology evaluation structure of the test chip can be divided into three parts: the degradation generation circuits, degradation indicators, and readout as shown in Fig. 4.1. The degradation generation circuits are built from the target technology and generate one or more degradation indicators, which can reflect the radiation sensitivity and severity. Then, these indicators are quantified by the measurement part. This section briefly discusses the state-of-the-art evaluation structure in both TID effects and SEEs. Then, the test of the Godzilla project is present for SEEs.

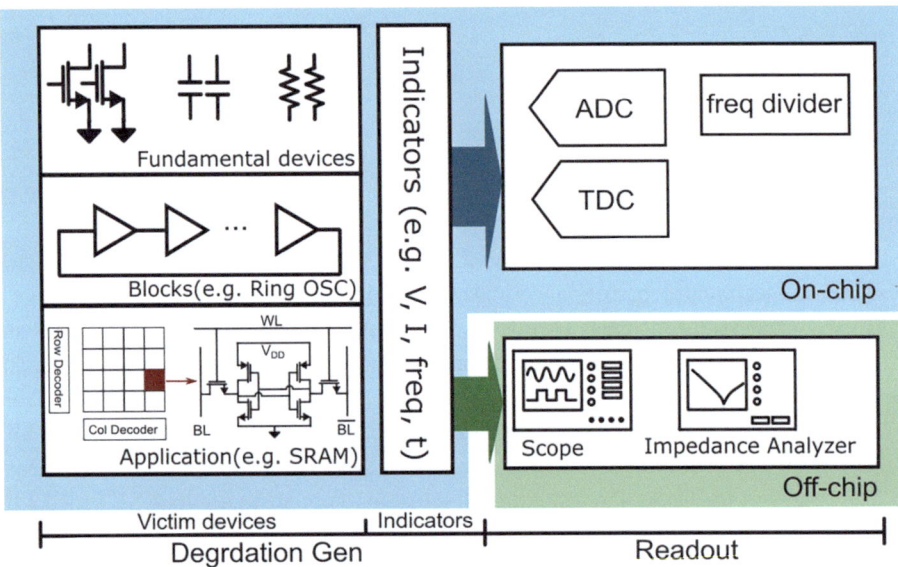

Fig. 4.1 Technology evaluation structure used in a test chip

4.2.1 Technology Degradation Generation Circuits

4.2.1.1 Victim Devices and Indicators

Technology degradation generation circuits are a series of dedicated circuits (victim devices) that are used to generate a certain type of indicator that can relate to radiation effects. These indicators can be divided into direct and indirect indicators. Direct indicators are generally related to and generated from the fundamental components of the technology, which include but are not limited to voltage, current, charge, time duration, time delay, and resonating frequency. Normally, these indicators can directly reflect the sensitivity or severity of the radiation effect in technology, etc. To produce these indicators, the degradation generation circuits can be the elementary components, such as transistors, resistors, and capacitors. From the response of the elementary components, an accurate radiation model can be built, and the upper-level application behavior under the radiation can be derived since they are composed of these fundamental elements. It needs to be noted that not all direct indicators

4.2 The Methodology of Technology Evaluation Under Radiation

can be acquired or measured conveniently and accurately. Thus, some indicators need to be converted to the other one to have a more efficient quantization. The list below details the features of the direct indicators and their generation circuits:

- **Voltage**: Voltage normally is one of the most often used direct indicators to characterize the radiation effects in a technology evaluation. Ideally, voltage change can be directly obtained when radiation effects happen to a single transistor. For example, the SET voltage fluctuation can be observed on the drain node of an off-state transistor with a high-speed digital oscilloscope [104]. Biased voltage change can be measured at the op-amp biasing circuits [105]. Output voltage of an inverter with a fixed input can also be used as an SET indicator since a temporary voltage fluctuation is generated when a SET happens [106]. However, some voltage indicators cannot be easily observed or easily affected by the measurement. One has to perform the indirect voltage measurement. For example, the threshold voltage shift due to TID effects is normally extracted by characterizing the I-V curves of a single transistor to derive the threshold voltage shift [107]. To matters worse, the electrostatic discharge (ESD) structure in the IO of the chip can disturb the accuracy of the leakage current measurement [108].
- **Current**: Similarly, current is another direct indicator that can express the radiation degradation. For example, in less advanced technology, TID effects can boost the single transistor leakage exponentially when the transistor is in an off state. DC current meters or Source Measure Units can be used to quantify this quiescent leakage current. As for dynamic current, such as SET current, it can also be quantified directly through an off-chip oscillator with an attenuation (bias-Tee) [109].
- **Charge**: In SEEs, energy particles can cause ionization in the semiconductor devices. In other words, deposited charge, or ionization charge, is generated in the circuits, which can also be measured and is also an important indicator to quantify the severity of SEEs. However, direct measurement of the ionized charge is difficult. Therefore, indirect measurement is often performed, such as integrating the SEE charge to a voltage [72] or converting the charge to a voltage drop and acquiring it by a peak detect and hold circuit [110].
- **Resonating frequency and Delay**: TID effects can cause threshold voltage shifts and intrinsic frequency changes in transistors. Therefore, the speed of circuits, like digital gates and the bandwidth of the amplifier, is also affected. Since the frequency and delay are correlated, both frequency and delay can be a useful indicator to evaluate the degradation of the technology. However, these two are not direct indicators since they can only be extracted through a certain type of circuit, such as ring-oscillators, op-amps. For example, as mentioned before, leakage current measurement can be a challenge due to the ESD structure at the TID investigation. As a result, the frequency degradation of the ring oscillator is used to derive the threshold voltage degradation [108].

Besides the direct indicator, indirect indicators are more often used when the technology and applications are closely bonded. In other words, these indicators partially depend on the technology and on the architecture/design of the applications.

- **Cross-section**: Cross-section refers to the probability of SEE happening, such as bit flip, latch-up, or other transient effects, when an energy particle interacts with a particular sensitive region of a device or circuits. More formally, the cross-section σ is defined as the number of events per incident particle fluence (number of particles per unit area), which can be expressed as Eq. (4.1). The typical units of cross-section are cm^2/bit (for memory devices) or cm^2/device. It represents the effective area of the device that is sensitive to particle interactions that can cause SEEs. It needs to be noted that cross-section is not a direct indicator of the technology evaluation since it relates to not only technology but also other factors, such as the architecture of the victim devices or circuits. Cross-section is not only limited to the SEEs with energy particles but is also used in the pulsed laser test,. In this case, the pulsed laser cross-section relates to the laser-scanned area S, the number of laser-scanned pixels, and the number of observed events and is expressed as (4.2) [111, 112].

$$\sigma_{particle} = No.Events/Fluence \qquad (4.1)$$

$$\sigma_{laser} = No.Events \cdot S/(No.pixels) \qquad (4.2)$$

- **Critical Charge/Energy**: Critical charge refers to the minimum amount of electrical charge that must be deposited by an energetic particle in a sensitive region of a semiconductor device to cause an SEE, such as a bit flip, latch-up, or other transient event. If the charge deposited by a particle exceeds this critical charge, the device will experience a state change or fault (e.g., a 0 changing to a 1 in a memory cell). Conversely, if the deposited charge is below the critical charge, the device remains unaffected. Since sufficient laser energy can also result in the ionization and also SEEs. Therefore, a similar definition of critical energy can be concluded for laser-induced SEEs [113].

4.2.1.2 Selection of Victim Devices

The victim devices refer to the test structures built by the target technology that we would like to explore for their radiation response and sensitivity. Victim devices can be built at different hierarchical levels, as shown in Fig. 4.1. The most simple and fundamental structure is the elementary devices, such as transistors, resistors, and capacitors.[1] Through the observation of fundamental structures, one can understand more from the physical aspects, and a device-level model can be built from the data. Beyond the fundamental device level, representative blocks can also be built to investigate a more application-related response. For example, ring-oscillators are widely used to investigate the speed and power degradation of the unit digital

---

[1] Passive devices are normally considered inherently ionizing radiation-tolerant [114]. However, recent studies demonstrate that passive devices, such as resistors, may be affected by the radiation-induced ionization [103].

4.2 The Methodology of Technology Evaluation Under Radiation

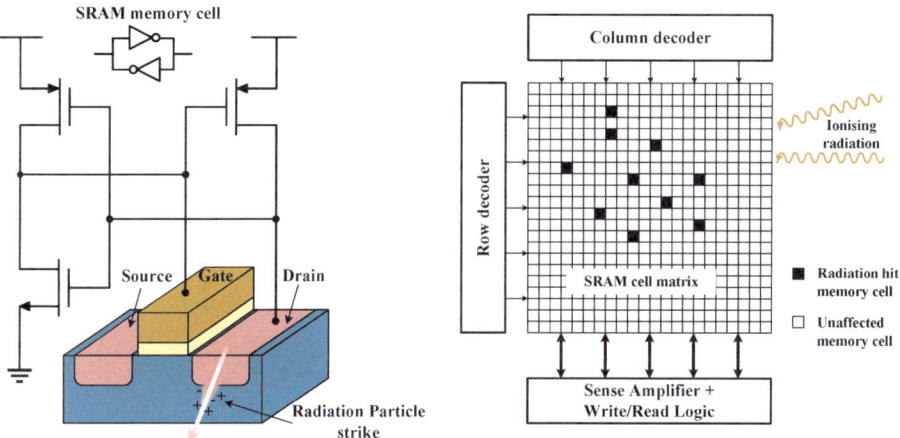

Fig. 4.2 Memory bit-flips on static random access memory (SRAM) nodes (*Source* Karmakar et al. [115])

cell since a tiny degradation in the transistor can reduce a significant frequency difference [108]. At an even higher level, some widely used application circuits are used as a benchmark to compare the technology sensitivity. For example, SRAM is widely used in space cache memory, control systems, and radiation sensors due to the speed and robustness [115, 116]. Therefore, SRAM, as shown in Fig. 4.2, can be a representative benchmarking circuitry and can be built to investigate and compare with different technologies. Besides, each technology contains several transistor flavors, which include but are not limited to core/IO devices (also referred to as different voltage domains), low/standard/high threshold voltage, devices w/wo DNW, etc. By implementing the same circuitry with different transistor flavors, one can also explore and compare the flavor effect on performance as well as radiation tolerance.

A special requirement for victim devices is that their area may directly affect the probability of SEE occurrence, in other words, the event frequency. For the SEE test under the particles like protons or heavy ions, the flux rate of the particle beam is limited. Therefore, to generate a sufficient observation (readout) in a limited time frame, the (sensitive) area of the victim devices must be large enough. The relation between the particle flux ϕ_{par}, the sensitive area A_{sens} and the event rate r_{event} can be expressed in

$$r_{event} = \frac{A_{sens}}{\phi_{par}} \quad \left[\frac{cm^2}{ions/(s \cdot cm^2)}\right] \quad (4.3)$$

It should be noted that a larger (sensitive) area results in a higher readout frequency, which will cause a higher data flux in the transmission to PC. Therefore, on-board intermediate storage may be required during the SEE test. On the other hand, a larger area can also result in a higher probability of a double-hit or even a multiple-hit event, which can also cause an inaccurate readout.

4.2.2 Readout

The readout unit is used to acquire and quantize the degradation indicators as shown in Fig. 4.1 and can be performed either off-chip or on-chip. Off-chip readout directly uses the equipment, such as an oscilloscope or FPGA, to acquire the degradation indicators. Since all signal acquisition, quantization, and processing are proceeding off-chip, the technology evaluation chip (or test vehicle) mainly consists of the victim devices and consumes less area and design effort compared to the one using on-chip readout. It also brings another advantage that there is no need to worry about the degradation of the measurement setup since they are normally placed outside of the radiation environment. However, it also results in several drawbacks. First, some measurements require a vacuum environment, and the test chip has to be placed inside of a vacuum chamber as shown in Fig. 4.3, which poses limitations to the signal inputs and outputs [117]. Besides, the large-scale characterization can be challenging since there is the limited number of IO pins. Though an on-chip multiplexer can help extend the accessibility, it can also cause worse measurement accuracy.

On-chip characterization is much more efficient and used in most of the technology evaluation chips. The radiation indicators are directly digitized on-chip by different kinds of converters, such as ADC, TDC, frequency dividers, etc. Then the digitized radiation indicators are transmitted in the digital domain, which improves the data rate and gives more reliability. This is an advantage for large-scale automatic characterization. However, the response of the on-chip readout circuit from the radiation effects must be considered. Circuit degradation may reduce the readout resolution, and the circuit error can cause the invalid data. This error may be temporary (e.g., SEU, MBU) or permanent (e.g., SEB, SEFI) [118, 119]. Therefore, the area of the readout circuit must be minimized to have a small cross-section. If the area is hard to minimize, then radiation hardening has to be considered. The hardening method can be referenced in the hardening method in Chap. 2. For example, using a thermometer code is an effective way to identify the SEU or MBU in the output codes. Figure 4.4 shows the normal and 4 error scenarios from SEEs of the thermometer

Fig. 4.3 Vacuum chamber used in heavy ion test at the heavy ion facility (HIF) in UC Louvain, Belgium

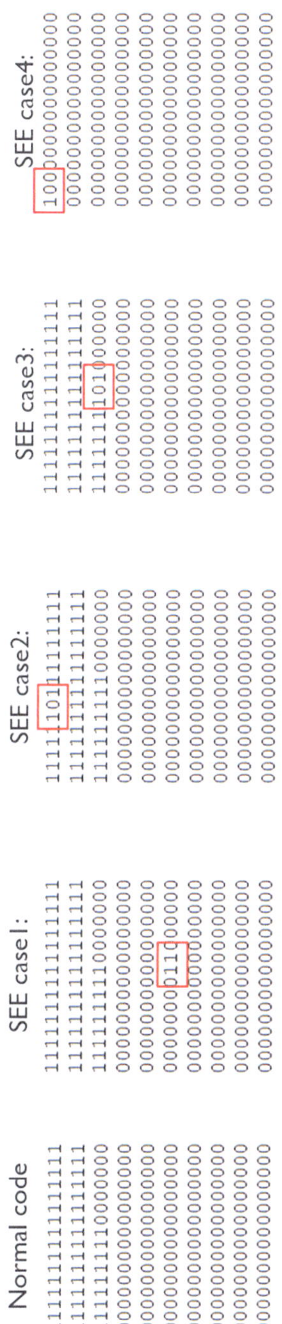

4.2 The Methodology of Technology Evaluation Under Radiation

Normal code	SEE case1:	SEE case2:	SEE case3:	SEE case4:
1111111111111111	1111111111111111	11111`101`1111111	1111111111111111	`100`0000000000000
1111111111111111	1111111111111111	1111111111111111	1111111111111111	0000000000000000
1111111110000000	1111111110000000	1111111110000000	111111`101`0000000	0000000000000000
0000000000000000	0000000000000000	0000000000000000	0000000000000000	0000000000000000
0000000000000000	00000000`111`00000	0000000000000000	0000000000000000	0000000000000000
0000000000000000	0000000000000000	0000000000000000	0000000000000000	0000000000000000
0000000000000000	0000000000000000	0000000000000000	0000000000000000	0000000000000000
0000000000000000	0000000000000000	0000000000000000	0000000000000000	0000000000000000

Fig. 4.4 Thermometer codes bring tolerance in SEEs, only case 3 and 4 cannot be calibrated and may cause limited readout error

codes. Cases 1 and 2 can be calibrated easily since the error is located in all 1 or all 0 streams. When the error happens at the 1-0 boundary or stream beginning, like in cases 3 and 4, calibration is hard to perform. However, these two cases are very rare, and the error amplitude is limited.

4.3 65-nm CMOS Technology Test Vehicle (Godzilla) for SET Evaluation

4.3.1 Position of the 65 nm CMOS Technology

CMOS is a type of MOSFET fabrication process that uses complementary and symmetrical pairs of P-type and N-type MOSFETs for electronic functions [120]. CMOS technology is widely used for constructing IC chips, including microprocessors, microcontrollers, memory chips, and other digital logic circuits. CMOS technology can also be used for analog and mixed-signal circuits such as image sensors, data converters, RF circuits, transceivers, and receivers for communication.

After the CMOS process was first conceived by Frank Wanlass at Fairchild Semiconductor and introduced by Wanlass and Chih-Tang Sah in 1963 [121], the CMOS technology is continuously scaled to achieve higher integration density, higher speed, and lower dynamic power dissipation. Figure 4.5 shows the CMOS technology roadmap till 2020 [122]. Over the past decade, the leading CMOS process nodes, including 22, 14, 10, 7, and 5 nm, have significantly reduced transistor size, enabling higher transistor density and improved power efficiency. These technologies are gradually becoming mainstream in advanced ASICs, microprocessors, and memory chips. However, the well-established CMOS technology nodes, such as 180, 90, and 65 nm, are still much favored by many applications. Till Q2 2022, as shown in Fig. 4.6, almost 46 % of the foundry revenue came from the $\geq$ 55 nm technology nodes [123]. If we further look at the market around the world in Fig. 4.7, $\geq$

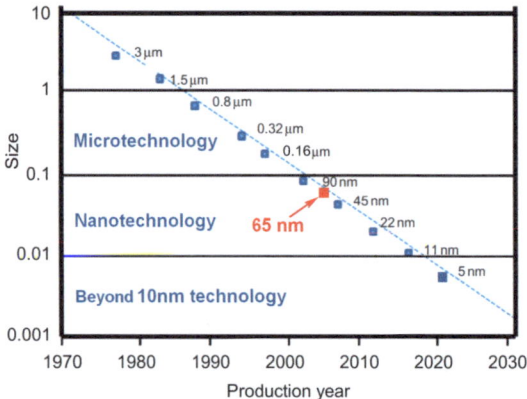

Fig. 4.5 Transistor gate length in technology nodes and production year

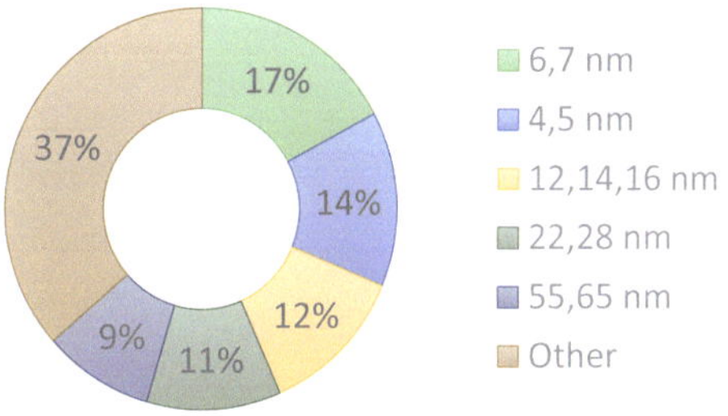

4.3 65-nm CMOS Technology Test Vehicle (Godzilla) for SET Evaluation

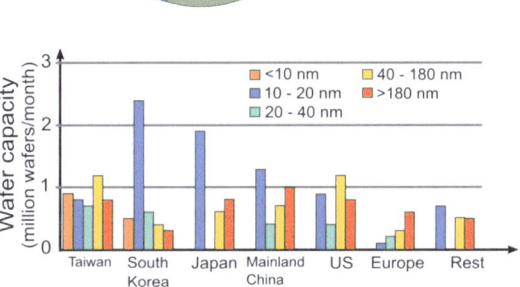

Fig. 4.6 Foundries revenue share of different technologies (*Source* Counterpoint [123])

Fig. 4.7 Market size of different technology nodes (*Source* Williams [124])

40 nm technology nodes are still the main players and have a comparable wafer capacity compared to the advanced nodes (< 22 nm) [124].

The main reasons for choosing the well-established, or "relatively less advanced," technologies are low power consumption, low design complexity, high reliability, and, most importantly, low cost. Figure 4.8 shows that the wafer costs for the 20 nm technology in 2018 are almost twice as high as those for 65 nm wafers[2] [125]. Besides the fabrication cost, the development cost increase is even more severe since the design complexity increase results in more complex rules, checks, EDA tools, and also asks for more powerful computation sources, as shown in Fig. 4.9. Normally, the increased cost is acceptable on an individual chip when the production volume is sufficiently large, which is the consuming electronic market, like cellphones and laptops. However, the market for radiation-hardened electronics has the features of high reliability, high customization, and long service life, but low production volumes. Therefore, blindly moving to the advanced technology nodes is not attractive for such chip providers. Thus, well-established technology, such as 65 nm technology, is one of the ideal candidates for radiation-hardened applications in space.

4.3.2 Status of the 65 nm CMOS Technology Evaluation Under Radiation

As discussed in the previous chapter, SEEs and TID effects pose the greatest risks to modern CMOS chips in radiation environments. Therefore, it is crucial to thoroughly investigate the

---

[2] Note that the figure is labeled revenue, but that is from the perspective of the foundry. To the foundry customer, it is a cost.

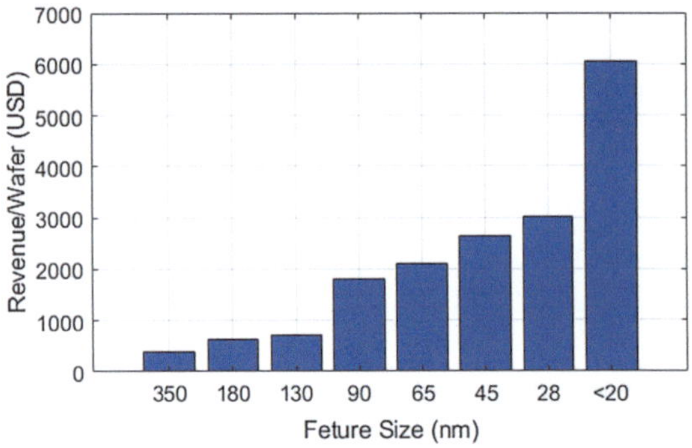

Fig. 4.8 Foundry revenue per logic wafer in the second quarter of 2018

Fig. 4.9 Cost of developing a new chip in different technology nodes (*Source* IBS 2014)

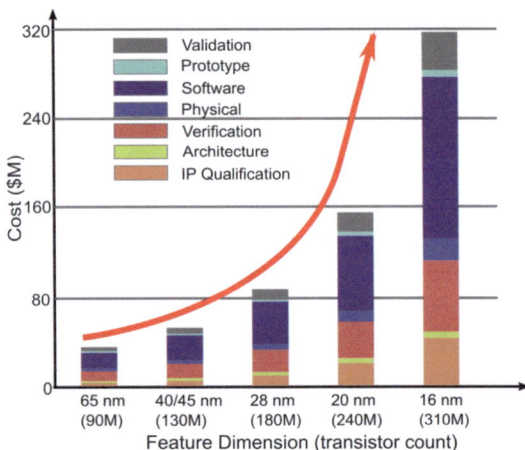

response of 65 nm technology to SEEs and TID before using it for radiation-hardened chip design. Typically, TID effects are examined by fabricating single-transistor structures with various dimensions and fundamental circuits on a test chip. Key electrical indicators such as threshold voltage and leakage current are measured before and after irradiation [126]. In a study by [127], TID effects on 65 nm technology were tested using an X-ray beam up to a dose of 200 Mrad.

For SEEs, the characterization of single-transistor SET responses provides data on ionized charge and current pulse characteristics, which are essential for developing accurate SET models for future transistor-level design. Various methods have been employed in previous research to characterize SET pulses [128]. In some studies, SET voltage pulses were measured directly using oscilloscopes [106, 129], though this approach has limitations due to parasitic effects and load influences. Other works have used latches with delayed signal paths to measure SET pulse widths on-chip [130], but this method requires multiple identical hits. Additionally, techniques involving a chain of identical cells as a delay chain have been employed to quantify transient pulse widths by counting the number of flipped cells

[131–134]. For 65 nm technology, several studies have used this on-chip method to measure SET pulse widths. However, the SET pulses were typically generated in inverter chains or sequential logic circuits [104, 135–137], which can be affected by Propagation-Induced Pulse Broadening (PIPB), leading to inaccurate evaluations [106, 138]. Furthermore, all such SET pulse measurements were performed at gate-level and the results were depends on the more than one transistors. Therefore, single transistor SET pulse measurement is still missing, which is very necessary for the SET model building. For charge measurement, previous research has only reported off-chip measurements of the collected charge [104], with no on-chip charge characterization performed for single transistors in earlier studies. By performing the on-chip SET pulse and charge measurement in one test chip, one can obtain sufficient information for the SEE model building.

4.3.3 The Scale of the Test Vehicle Godzilla

This test vehicle was developed to assess the SET effects in a 65 nm CMOS technology. The measurement circuits are designed to evaluate two key parameters: total SET ionization charge and SET pulse duration, both of which are measured directly on-chip. Various transistors, differing in type and size, were used as victim devices to examine how these factors influence SET effects. The variation in SET behavior with changes in supply voltage was also explored. Testing was conducted under a heavy-ion beam with effective LET values ranging from 20.4 to 88.35 MeV·cm^2/mg, using incidence angles from 0° to 45°. The SET characterization results are detailed present in later sections.

4.4 Target Devices

4.4.1 Target Devices Sensitive Area Calculation

The reverse-biased junction is the most charge-sensitive part of a MOSFET. The SET worst case happens when a MOSFET is in off-state, and an energetic particle (proton or heavy ion) passes the drain depletion region (Fig. 4.10). Then, a voltage pulse will appear at the drain node [139]. This voltage pulse can be used to quantify the total ionization charge and pulse duration by different measurement circuits. When calculating the sensitive area of the test chip and the hit probability, only the drain size is considered. However, from the SET results demonstrated in the later part of this chapter, the gate and even the source area also need to be considered. This is also confirmed by the SEE trend in Section 2.6.3. Since the heavy-ion beam can only reach a certain flux, this means the sensitive area needs to be large enough to have a high enough SET probability. However, a larger sensitive area means a larger chip area and tapeout costs.

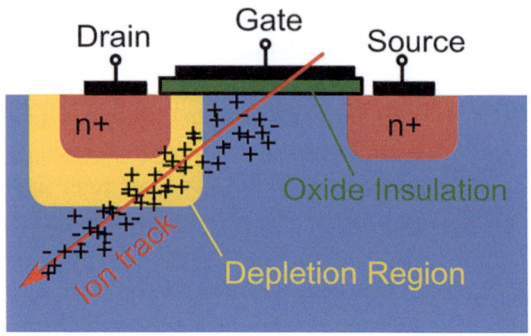

Fig. 4.10 SET causes ionized electrons and holes in off-state NMOS

To get sufficient readouts (hits) to the victim devices, the sensitive area (total drain area) needs to be large enough. The maximum heavy ion flux at the Heavy Ion Facility in Université Catholique de Louvain, Belgium is 15k ions/(s·cm^2). In this design, the hit probability chosen to be at least 1 hit/3 s (100 hits/5 min) to leave some readout flexibility. Then, the sensitive area for each type of device needs to be:

$$A_{sens} = \frac{1 hit/3\,s}{10 kions/(s \cdot cm^2)} = 3333 \mu m^2 \qquad (4.4)$$

Besides enough sensitive areas, double hits measurement also needs to be avoided. The measurement data is read every 10 ms (100 Hz) to avoid double hits. The probability of double hits in 10 ms can be calculated by the Poisson distribution (λ = 1/3/100 hit/period, k = 2):

$$P(k=2, 10ms) = \frac{\lambda^k e^{-\lambda}}{k!} = 5.55 \times 10^{-6} \qquad (4.5)$$

From the result, the double hits probability is low, so the situation that more than two hits in 10 ms can be ignored. As a consequence, there are only three situations: zero, single hit and double hit. If the experiment time is 1 min, the probability of a double hit is:

$$P(k=2, 1min) = 1 - [1 - P(k=2, 10ms)]^{10/0.01} = 0.0328 \qquad (4.6)$$

On the other hand, the probability of zero hits in 1 min can also be calculated:

$$P(k=0, 1min) = [1 - P(k=1, 10ms) - P(k=2, 10ms)]^{10/0.01} = 1.9 \times 10^{-9} \qquad (4.7)$$

4.4.2 Selected Victim Devices

In 65 nm technology, there are more than ten types transistors that the designer can freely use. All of these transistors can be categorized differently in different criteria. In different voltage domain, these transistors can be categorized in IO transistors (1.8, 2.5, 3.3 V) and core transistors (1.2 V). By different threshold voltage, transistors can be divided into low-threshold (lvt), medium-low-threshold (mlvt), standard-threshold (svt) and high-threshold (hvt) transistors. However, if the test chip covers all transistor types with different dimensions, the area can be significant large. Due to this, eight typical MOSFETs were chosen as victim devices and implemented. These transistors are most often used by most digital and analog circuits. The size information of the victim devices is shown in Table 4.1. Each type of victim device has a total sensitive area of 3333 μm^2. Several comparisons can be made between different sizes and types of victim devices:

- Characterizing SET between N-type and P-type devices.
- Characterizing SET between DNW and non-DNW devices.
- Characterizing SET between L = 60 nm and L = 500 nm.
- Characterizing SET between Core devices (thin oxide transistors with 1.2 V normal supply voltage) and IO (thick oxide transistors with 2.5 V normal supply voltage) devices.
- Characterizing SET between the different power supplies.

Table 4.1 Target devices implemented on chip

Index	Victim devices	Width μm	Finger	Length (nm)
VD0	Core NMOS	10	10	60
VD1	Core NMOS	10	10	500
VD2	Core DNW NMOS	10	10	500
VD3	Core PMOS	10	10	60
VD4	Core PMOS	10	10	500
VD5	IO NMOS	10	10	500
VD6	IO PMOS	10	10	500
VD7	IO DNW NMOS	10	10	500

4.5 Ionization Charge Measurement Circuits

4.5.1 Main Measurement Circuits

The ionization charge collected during a SET event can be determined by integrating the voltage pulse observed at the drain node [120]. The block diagram for SET charge measurement, which uses charge integration for N-type victim devices, is shown in Fig. 4.11. The measurement circuit for P-type devices follows a similar structure.

To establish a high-resistance node at V_{SET}, N off-state NMOS victim devices are biased using a resistor $R_{bias} = 1\,M\Omega$. The value of N is calculated based on the expected hit probability and the sensitive drain area of the MOSFET, as previously discussed. Additionally, the parasitic capacitance at V_{SET} is set to 5 pF to ensure uniform loading across all victim devices. The bias voltage $V_{bias,vd}$ applied to the victim devices can be adjusted, allowing for the evaluation of how supply voltage influences the SET charge characteristics. When an SET occurs in one of the victim devices, the resulting voltage pulse at V_{SET} is integrated by a voltage integrator. The integrator uses a feedback loop with five binary-weighted capacitors controlled by digital input bits $SW_{cap}[4:0]$. By adjusting $SW_{cap}[4:0]$, the integrator output V_{out} can be modified according to Eq. (4.8).

$$V_{out} = \frac{Q_{SET}}{C_{unit} n_{cap}}
= \frac{1}{C_{unit} \sum_0^4 2^i SW_{cap}[i]} \int_0^t \frac{V_{SET}}{R_{int}} dt \qquad (4.8)$$

where Q_{SET} is the ionization charge from the SET at node V_{SET} and n_{cap} is the number of activated unit feedback capacitors in the integrator.

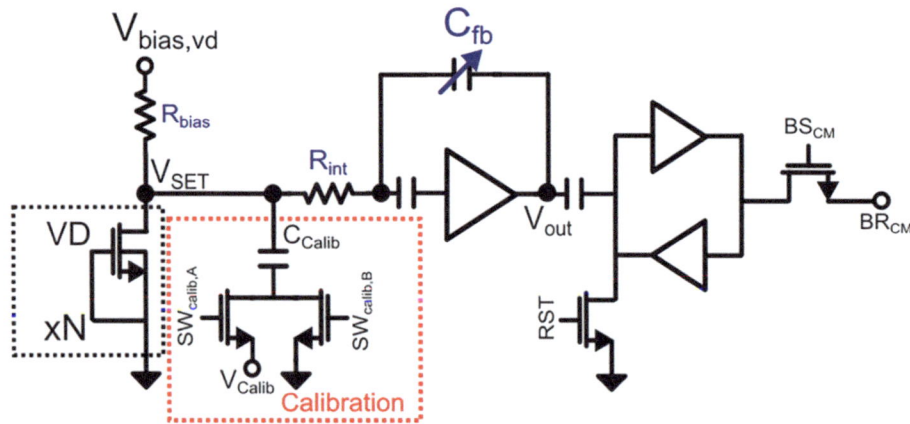

Fig. 4.11 Block diagram of the N-type SET charge measurement

4.5 Ionization Charge Measurement Circuits

After integration, the output voltage is fed into a latch. The latch input is reset to logic low before each measurement. If no SET occurs or if V_{out} is below the latch threshold $V_{th,latch}$, the output BR_{CM} remains low. However, when the latch threshold is exceeded, BR_{CM} transitions to high. By adjusting the number of connected capacitors, the threshold charge Q_{th} of the measurement circuit can be configured as Equation (4.9). If the collected charge Q_{SET} surpasses Q_{th}, BR_{CM} will be set to high.

$$Q_{th} = V_{th,latch} C_{unit} \sum_{0}^{4} 2^{i} SW_{cap}[i] \tag{4.9}$$

4.5.2 Charge Measurement Calibration

Prior to the irradiation test, it is necessary to characterize the transfer function between the threshold charge Q_{th} and the number of capacitors n_{cap}. This is done by injecting a known charge into the integrator. Referring to Fig. 4.11, capacitor C_{calib} is connected to the node V_{SET}. The value of C_{calib} is measured using a network analyzer. During calibration, $SW_{calib,A}$ is closed, and $SW_{calib,B}$ is open, allowing C_{calib} to be connected to the calibration voltage V_{calib}. Once the circuits are fully reset, $SW_{calib,A}$ is opened, and $SW_{calib,B}$ is closed. Since V_{SET} is a high impedance node, the injected charge is given by $V_{calib} C_{calib}$. To calibrate the threshold for a specific capacitor setting, V_{calib} is swept from low to high. The threshold charge Q_{th} corresponds to $V_{calib} C_{calib}$ at the point where BR_{CM} switches states.

In total, there are eight types of victim devices, each equipped with its own measurement circuit to eliminate parasitic effects from interconnections. All measurement circuits are calibrated separately before the irradiation test. As an example, the calibration results for Core NMOS with $L = 60$ nm are illustrated in Fig. 4.12. The dashed and solid lines represent the simulation and calibration results, respectively. The y-axis indicates the number of capacitors n_{cap}. From the results, n_{cap} exhibits a staircase pattern as V_{calib} increases. The threshold charge Q_{th} for each capacitor level is determined by the transition point of the staircase and is derived from Eq. (4.9).

Due to slight variations in parasitic capacitance at V_{SET}, each transistor type's measurement circuit demonstrates a different measurement range. For instance, the measurement circuit for the core NMOS victim devices with $L = 60$ nm can detect ionization charges ranging from 188.24 fC to 1.41 pC, with an average resolution of 40.61 fC/step. The overall calibration results for the eight victim devices are summarized in Table 4.2. The minimum and maximum threshold charges, $Q_{th,min}$ and $Q_{th,max}$, indicate the lowest and highest detectable charge values, respectively. Q_{step} represents the average charge per capacitor level in the transfer function.

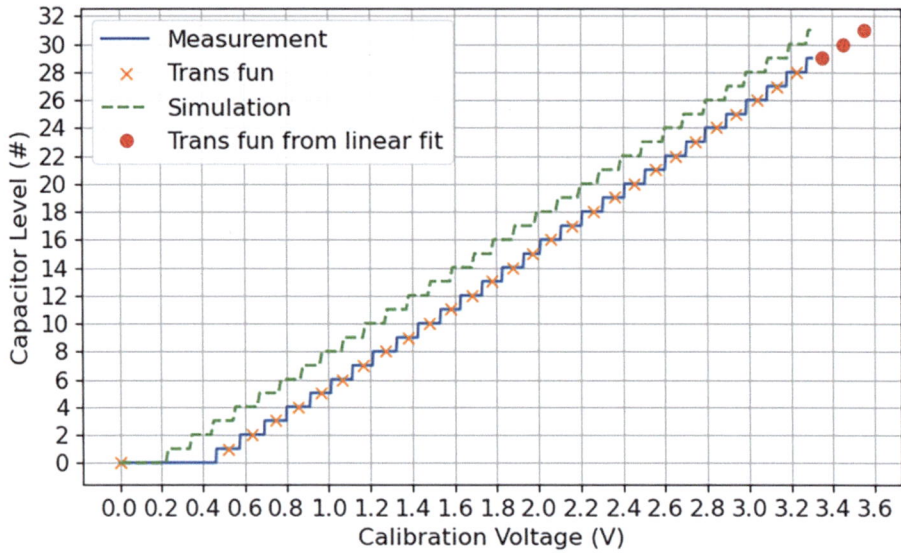

Fig. 4.12 Core NMOS victim devices with L = 60 nm calibration results

Table 4.2 Entire calibration results of eight victim devices (W= 10 μm, F= 10)

Victim device	$Q_{th,\min}$ (fC)	$Q_{th,\max}$ (pC)	Q_{step} (fC/level)
Core NMOS L = 60 nm	188.24	1.41	40.61
Core NMOS L = 500 nm	177.63	1.40	40.69
Core DNW NMOS L = 500 nm	162.90	1.39	40.63
Core PMOS L = 60 nm	212.21	1.65	47.63
Core PMOS L = 500 nm	206.63	1.63	47.30
IO NMOS L = 500 nm	274.11	3.81	117.46
IO PMOS L = 500 nm	321.88	4.33	133.24
IO DNW NMOS L = 500 nm	258.73	3.75	115.98

4.6 Pulse Duration Measurement Circuits

4.6.1 Main Measurement Circuits

The pulse duration measurement circuits consist of three main components: SET pulse generation blocks, a combiner, and a Memory Delay Line (MDL). The structure of the pulse duration measurement circuits for N-type victim devices is illustrated in Fig. 4.13, while the circuits for P-type victim devices follow a similar design.

For each type of victim device, N devices are distributed across M SET pulse generation blocks. In each block, N/M off-state victim devices are biased by an on-state PMOS transistor $M1$. When an ionizing particle strikes one of the victim devices, the resulting SET voltage pulse at node V_{SET} is amplified into a square wave by amplifier AMP1. The reason for splitting the victim devices into multiple blocks is that SET pulses are sensitive to the time constant at V_{SET}. By using M SET generation blocks, the parasitic capacitance at node V_{SET} is kept below 100 fF. The advantage of using pulse generation blocks, compared to the conventional inverter chain method described in [104, 135], is that the SET voltage pulse is delivered directly to the measurement circuits without passing through additional inverter stages. This approach helps to avoid side effects such as pulse broadening.

Since the victim devices are distributed across M pulse generation blocks, an M-input combiner is used to collect the SET pulses and transmit them to the MDL. The combiner has M inputs and is composed of four stages of 4-input NOR and NAND gates. The delay between the inputs and outputs is crucial for maintaining measurement accuracy, and minimizing delay differences is necessary to prevent signal distortion. To achieve this, balanced NOR and NAND gates are employed. Figure 4.14 illustrates the balanced NAND gate (the same principle applies to the NOR gate). Redundant NMOS transistors are added to the lower part of the NAND gate to equalize parasitic effects for each input. Additionally, the gates are distributed among the victim devices in a tree structure, as depicted in Fig. 4.15.

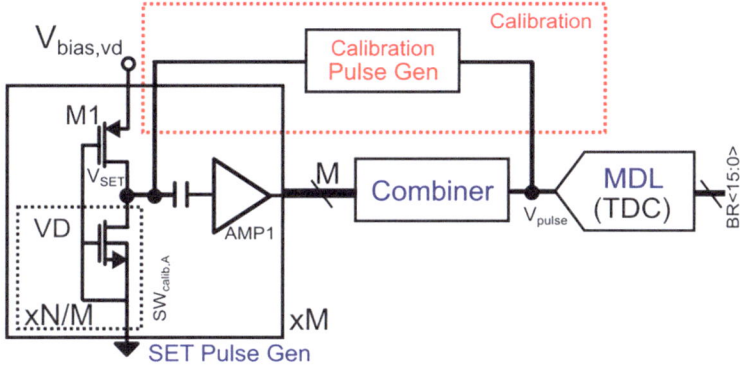

Fig. 4.13 Block diagram of the N-type SET pulse duration measurement

Fig. 4.14 Balanced NAND gate used in the combiner

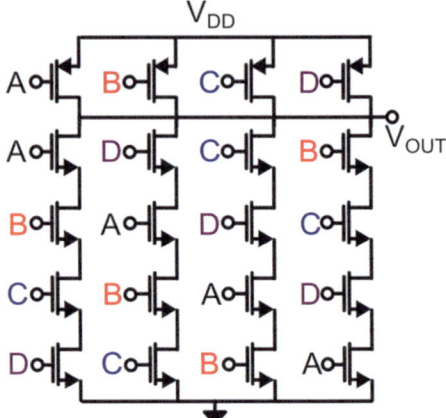

Fig. 4.15 SET pulse combiner collection tree

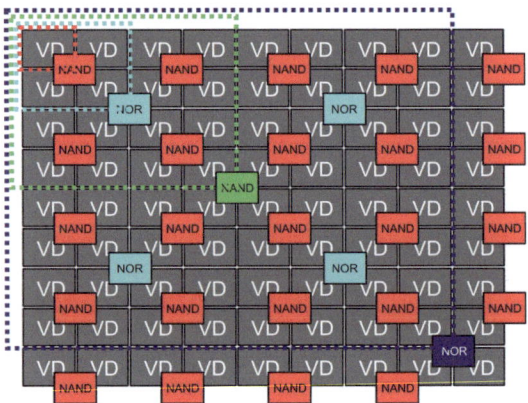

Similar to a clock distribution tree, the wire lengths from each victim device to the output are kept identical. As a result, the simulated maximum delay difference after layout is less than 6 ps.

The MDL, illustrated in Fig. 4.16, uses inverter delay as a reference to measure the pulse duration. It consists of 160 Memory Delay Cells (MDCs), where each MDC contains a latch and a switch. The input switch of each MDC is controlled by the SET pulse signal V_{pulse}, originating from the combiner output. Prior to the pulse duration measurement, the inputs of the odd and even MDC stages are reset to high and low, respectively.

When the pulse signal arrives (i.e., V_{pulse} transitions from high to low), the switches between the latches close, allowing V_{SS} to propagate through the MDL. Once the pulse ends, the switches open, stopping the propagation of V_{SS}. The average stage delay of the MDL is optimized by alternating low-threshold PMOS and NMOS inverters in the MDCs. The pulse duration is calculated by multiplying the number of flipped cells by the average inverter delay. Additionally, the measurement circuits (bias transistor $M1$, amplifier $AMP1$,

4.6 Pulse Duration Measurement Circuits

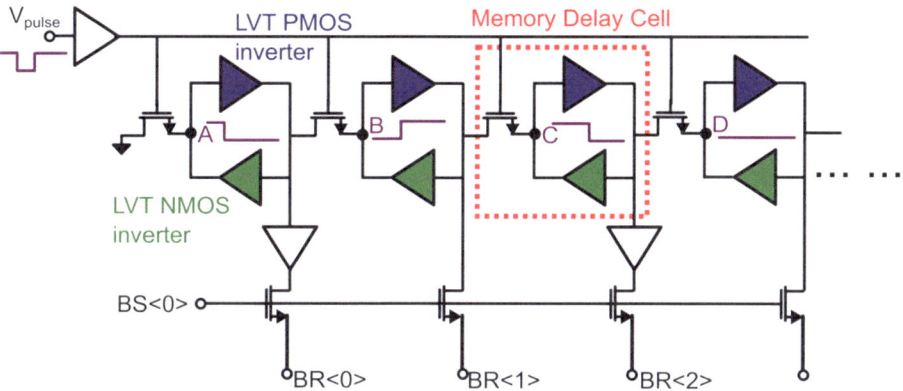

Fig. 4.16 Block diagram of the memory delay line (MDL)

combiner, and MDL) have a significantly smaller sensitive area compared to the victim devices. These circuits are implemented with a deep N-well (DNW) structure to prevent SET charge sharing.

4.6.2 Pulse Measurement Calibration

Before conducting the irradiation test, the pulse duration circuit is calibrated using a known pulse generated by the pulse generation circuit (shown in Fig. 4.17). This pulse generation circuit is built from a 1601-stage configurable ring oscillator. The large number of stages was selected to minimize the limited load effect that could occur on the outputs of certain

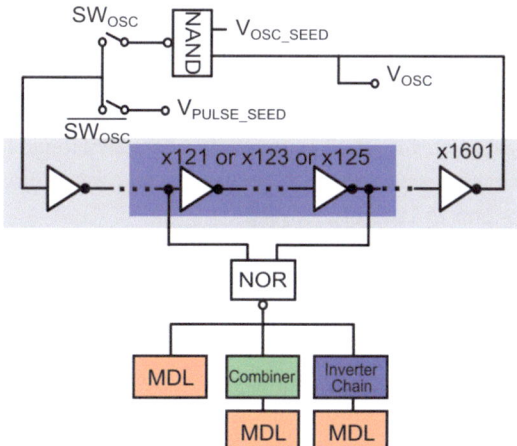

Fig. 4.17 Pulse generation circuit for calibration

Table 4.3 The calibration results of the MDL and capture circuits

MDL accuracy	19.42 ps/step
MDL detection threshold	38.9 ps
MDL range	3.14 ns
N-type capture circuit delay	233 ps
P-type capture circuit delay	252 ps

inverters. The oscillator operates in two modes: closed-loop ($SW_{OSC} = 1$) and open-loop ($SW_{OSC} = 0$). In the closed-loop mode, the average delay of the inverter, $T_{inv,osc}$, in the oscillator can be determined by measuring the oscillation frequency via V_{OSC}, using the relation in Eq. (4.10).

$$T_{inv,osc} = \frac{1}{f_{osc,out}(1601 \times 2)} \quad (4.10)$$

where $f_{osc,out}$ is the oscillating signal frequency at $V_{osc,out}$. The oscillating signal is generated by setting $SW_{OSC} = 1$ and switching $V_{OSC_{SEED}}$ from low to high.

Next, the oscillation loop is cut off by switching SW_{OSC} from high to low, and a 1600-stage inverter chain is formed. A pulse with the duration of $T_{inv,osc} \cdot m$ can be generated by connecting an m-stage (m = 121, 123 or 125) inverter chain's input and output with a NOR gate. In this way, a known duration pulse with a different duration is generated by changing V_{PULSE_SEED} from high to low. This known pulse is first sent to the MDL to characterize the average stage delay of the MDC. Then, an identical pulse is sent to the node V_{SET} in Fig. 4.13. The readout difference of the MDL is the delay of the pulse capture circuit (the amplifier AMP1 and combiner), which only acts as an offset to the SET duration measurement results,

The 1601-stage re-configurable ring oscillator, which is shown in Fig. 4.17 is first closed by SW_{osc}. From the measured oscillation frequency at $V_{osc,out}$ and Eq. (4.10), three pulses are generated: 1.184, 1.4195 and 1.767 ns. These pulses are sent to the pulse measurement circuit. The results are shown in Table 4.3. The MDL is characterized first, and the average stage delay is 19.42 ps/step with a detection range from 38.9 ps to 3.14 ns. Then, the capture circuit delay of the N-type and P-type capture circuits are derived, and the results are 233 ps and 252 ps, respectively.

4.7 Test Chip and Test Setup

The test chip was manufactured in a commercial 65 nm technology, and the dimension of the die is 2.89 mm × 2.33 mm. The die photo is shown in Fig. 4.18a. Instead of using a single complete IO ring, a staggered (two rows of IO on each chip side) two-sided IO form

4.8 SET Experimental Results

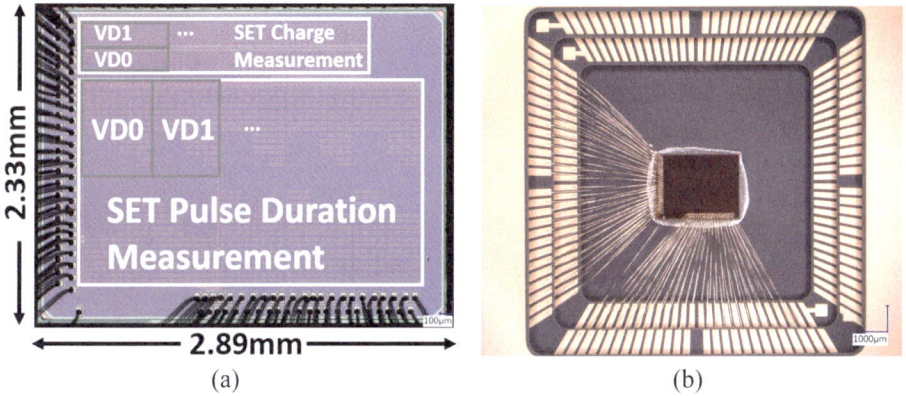

Fig. 4.18 Die and package photo of the 65 nm test vehicle: **a** die photo **b** package photo

is used to reduce the chip area. Since the IOs are mostly low-speed digital signals, bond wire parasitics can be ignored. A 256-pin ceramic pin-grid array package is used for the packaging. The package (shown in Fig. 4.18b) has a window on top, from which the heavy ion beam can access the die directly without any thinning.

The experimental setup is divided into two parts: the main test board and supporting boards. The main test board, shown in Fig. 4.19a, contains the test chip, which is also called DUT (Device Under Test), and will be put in the cyclotron vacuum chamber. The supporting boards of the experimental setup, shown in Fig. 4.19b, are used to assist the main test board in completing the SET experiments. They are made up of a connection board, a power manager board, an FPGA board and a USB 3.0 connector. These boards will be placed outside the cyclotron vacuum chamber. The connection board is located between the main test board and the FPGA board. The Low-dropout regulators (LDOs) on the connection board generate all the Core/IO supplies for the test chip. It also adapts the FMC connector (used by the FPGA board) to the D-SUB25 connector (on the main board) for digital input/output signals. During the SET test, all the power supplies can be monitored or changed by a power manager board in real time. When the SET experiment is finished, the digital data will be saved in the memory of the FPGA board and can be read by a PC through the USB 3.0 Connector.

4.8 SET Experimental Results

4.8.1 Test Chips and Heavy Ion Test Conditions

The test chips were subjected to irradiation using heavy ions at the Heavy Ion Facility (HIF) in UCLouvain, Belgium (as shown in Fig. 4.20). Two chips were tested under room temperature conditions, both displaying consistent results. Throughout the experiment, over 20,000 SET events were recorded for each chip. The irradiation parameters are listed in

Fig. 4.19 The experimental setup: **a** the main test board will be put in the cyclotron vacuum chamber. **b** The supporting boards for the SET experiments

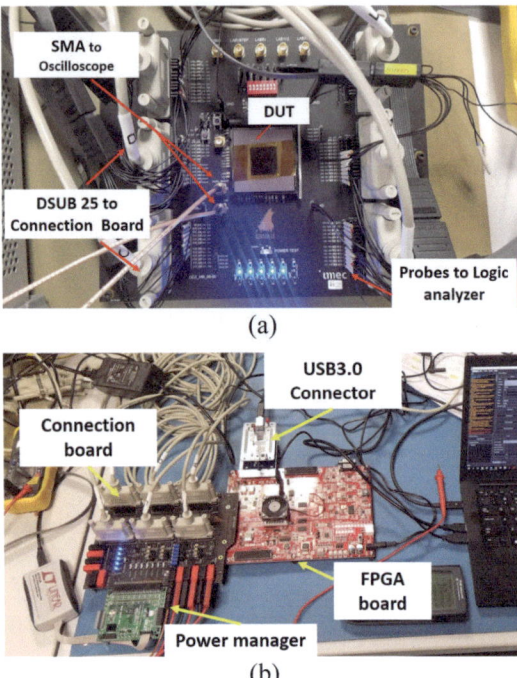

Table 4.4. In this test, Nickel and Xenon ions were utilized, with incidence angles of 0° and 45°, respectively, to achieve varying effective LET values. Additional details regarding the HIF parameters and the particle characteristics are available in [140].

4.8.2 Ionization Charge Measurement Results

Since the measurement circuits do not directly quantify the ionization charge, they instead compare Q_{SET} to Q_{th}. Therefore, for each irradiation condition, measurements were performed multiple times with different Q_{th} values to obtain a comprehensive distribution. In every measurement, the same effective fluence $Fluence_{eff} = 1.5 \times 10^6$ ions/cm² was achieved by adjusting the heavy ion flux and test duration.

4.8.2.1 Core Devices' Measurement Results

As indicated in Table 4.2, the chip incorporates three types of core NMOS transistors and two types of core PMOS transistors. The measurement results for the minimum length (60 nm) core NMOS and PMOS transistors are displayed in Fig. 4.21. When the SET ionization charge exceeds Q_{th}, a reading is registered at the output, representing one occurrence. At

4.8 SET Experimental Results

Fig. 4.20 Test campaign performed at the heavy ion facility (HIF) in UCLouvain, Belgium

Table 4.4 Irradiation conditions used during the test

Ion	LET (MeV · cm^2/mg)	Angle	LET$_{eff}$ (MeV · cm^2/mg)
Nickel	20.4	0°	20.4
Nickel	20.4	45°	28.9
Xenon	62.5	0°	62.5
Xenon	62.5	45°	88.4

the maximum effective LET of 88.39 MeV·cm^2/mg, the NMOS and PMOS devices record 277 and 279 occurrences, respectively, at the minimum Q_{th}. As Q_{th} increases, the number of occurrences decreases. The occurrences at the minimum Q_{th} are used to determine the unit device cross-section from Eq. (4.11). The calculated unit cross-sections for the NMOS and PMOS transistors are 4.85× 10^{-7} cm^2 and 4.88× 10^{-7} cm^2, respectively. However, as Q_{th} increases, the number of SETs decreases more rapidly for the PMOS transistors compared to the NMOS transistors. The maximum collected charge is measured when the number of occurrences approaches zero. The maximum charges for the PMOS and NMOS transistors are 1.274 pC and 0.898 pC, respectively. At lower effective LET levels, both the maximum collected charge and the unit device cross-section for the PMOS transistors are lower than those of the NMOS transistors. At the lowest effective LET, the critical charge of the PMOS at node V_{SET} does not reach the threshold required to trigger the latch, even at the minimum Q_{th}.

$$CS_{unit,CM} = \frac{Occurrences @ Q_{th,min}}{Fluence_{eff} \cdot No.Devices} \qquad (4.11)$$

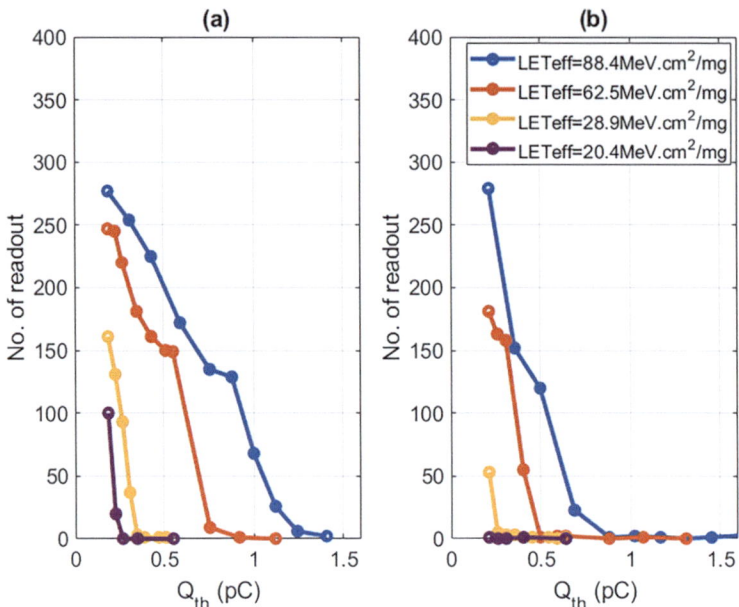

Fig. 4.21 Measurement results of core **a** NMOS and **b** PMOS with L = 60 nm and 1.2 V supply voltage

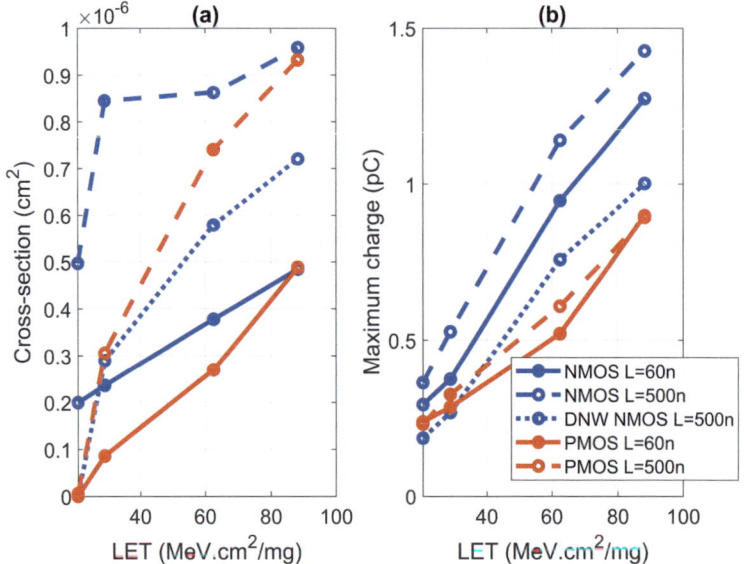

Fig. 4.22 Core victim devices measurement results with 1.2 V supply voltage: **a** unit device cross-section, **b** maximum collected charge

Similar assessments are conducted for all core victim devices. Figure 4.22a displays the unit device cross-sections and maximum collected charges of all victim devices across various effective LET values. As the effective LET increases, the unit device cross-sections for the core devices also rise, showing a tendency to converge at high effective LET values. Generally, the unit device cross-section exceeds the anticipated unit device drain area, necessitating consideration of the gate and source areas as well.

When heavy ions strike areas other than the reverse-biased drain junction, such as the substrate, the diffusion of charge within the silicon substrate or local p/n-well regions also contributes to charge collection [128]. The extent of diffusion charge is influenced by the ionization radius of the heavy ions in silicon and the distance from the impact site to the drain node. According to [141], 210 MeV Cl ions (LET = 11.4 MeV·cm^2/mg) possess an ionization radius of up to 1 μm in silicon. Consequently, if the distance from the impact point to the drain is less than this ionization radius, charge can be collected through diffusion, as well as by drift for charges outside the funnel that enter the drain depletion region. This is corroborated by results from devices with varying gate lengths but identical drain areas, where devices with longer gate lengths exhibit larger unit device cross-sections, as shown in Fig. 4.22a.

Additionally, it is noted that PMOS devices display a smaller or equal unit device cross-section compared to NMOS devices across all effective LET levels. For DNW NMOS devices, the unit device cross-section is diminished relative to non-DNW NMOS devices. This reduction occurs because some of the charge generated in the p-well region is collected by the DNW before it can reach the NMOS drain [142]. Furthermore, charge located beneath the p-well/DNW junction cannot be collected by the NMOS drain due to the triple-well collection toward the n-well tap [142, 143].

Figure 4.22b illustrates the maximum collected charge for all victim devices at various effective LET levels. The data indicates that the maximum collected charge increases with effective LET; however, no convergence tendency is observed. Consistent with the findings regarding unit device cross-sections, PMOS devices demonstrate a smaller collected charge compared to NMOS devices, aligning with off-chip charge measurements reported in [104]. Furthermore, devices with longer channel lengths exhibit a greater amount of collected charge.

4.8.2.2 IO Devices' Measurement Results

This test chip also includes an examination of three types of IO devices. The same methodology was applied to calculate both the unit device cross-section and the maximum collected charge, which are illustrated in Fig. 4.23a and b, respectively. Similar trends observed in the core devices are evident here as well. Notably, the maximum collected charge for IO devices is greater than that for core devices when both have a channel length of 500 nm. This can be attributed to the higher bias voltage applied to the IO devices, resulting in a thicker

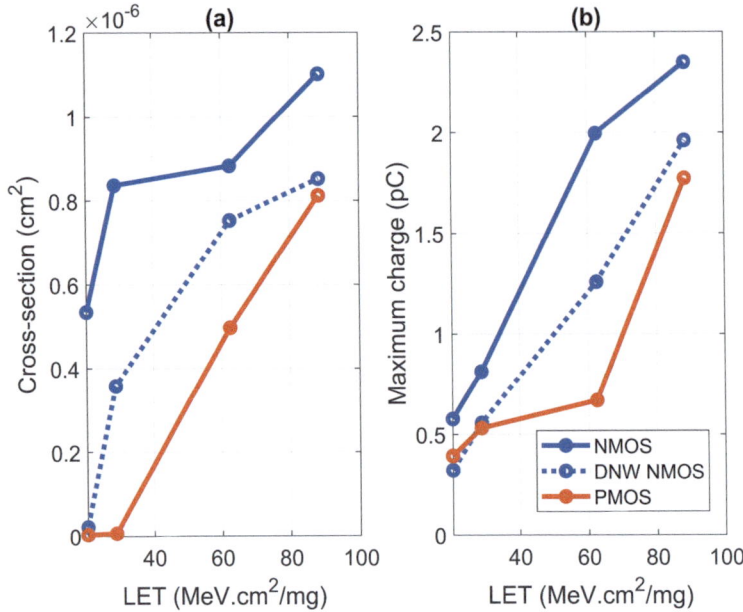

Fig. 4.23 IO victim devices measurement results with 3.3 V supply voltage and 500 nm channel length: **a** unit device cross-section, **b** maximum collected charge

depletion region due to the increased reverse bias voltage. Consequently, a larger portion of the ionized charge resides within the depletion region and is effectively collected [144].

4.8.2.3 Measurement Results with Different Supply Voltage

To investigate the effects of supply voltage on the collection of SET ionization charges, various supply voltages were applied to both core and IO devices. For the core devices, supply voltages of 1.08, 1.2, and 1.32 V were utilized. The results, depicted in Fig. 4.24a, indicate that the curves corresponding to different supply voltages are closely aligned, showing no significant differences. For the IO devices, results at supply voltages of 1.8, 2.5, and 3.3 V are presented in Fig. 4.24b. With the exception of the minimum Q_{th}, the occurrences of identical Q_{th} values increase with higher supply voltages. Furthermore, the maximum collected charge also rises with increased supply voltage, supporting the expectation that a higher supply voltage leads to a greater collection of charge for the same transistor.

4.8 SET Experimental Results

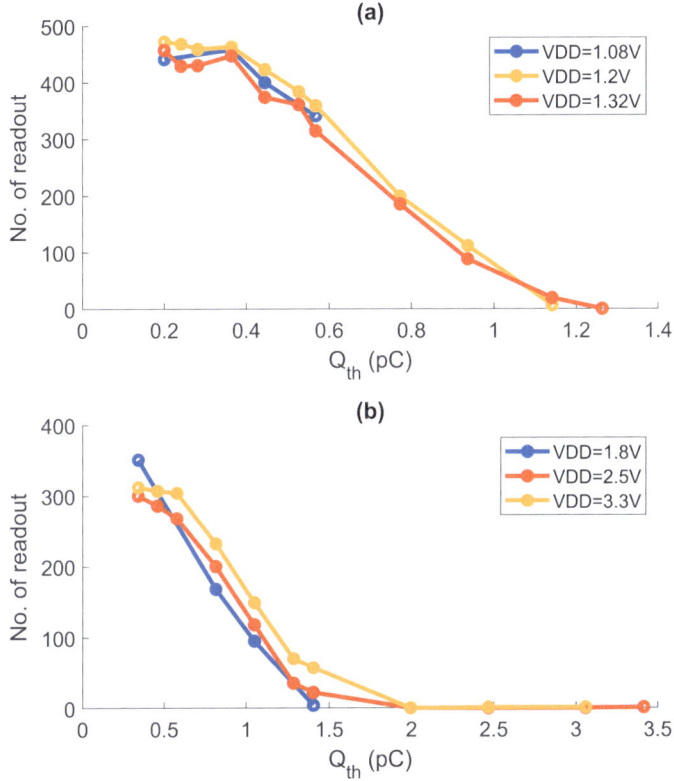

Fig. 4.24 Charge measurement versus different supply voltage: **a** core NMOS with VDD = 1.08 V, 1.2 V, 1.32 V, **b** IO NMOS with VDD = 1.8 V, 2.5 V, 3.3 V

4.8.3 Pulse Duration Measurement Results

Using the pulse duration measurement methodology, the distribution of pulse durations can be directly obtained from the readout. The pulse duration measurements for all eight types of victim devices were conducted under the same effective fluence, $Fluence_{eff} = 6 \times 10^6$ ions/cm^2, and the results were compared. In this measurement, transistors are utilized instead of resistors to adjust the biasing current of the victim devices. Consequently, N-type and P-type victim devices exhibit different recovery currents when an SET occurs, complicating a fair comparison of pulse durations between the two types. Therefore, the duration measurement results are compared exclusively within the same type of victim device.

The results of SET pulse duration measurements for core NMOS transistors with channel lengths of 60 and 500 nm, as well as for core DNW NMOS transistors with a channel length of 500 nm, are presented in Fig. 4.25a–d. Each SET pulse measurement is counted

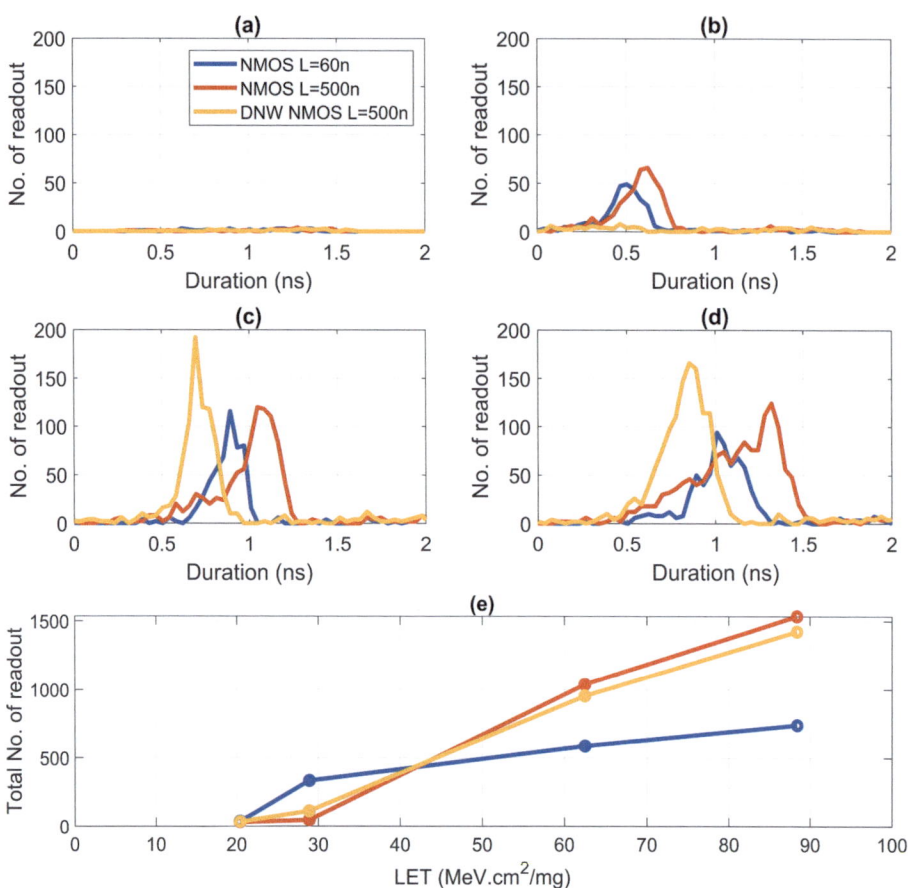

Fig. 4.25 The pulse duration distribution of core NMOS, DNW NMOS and PMOS transistors with 1.2 V supply voltage at LET of **a** 20.4, **b** 28.9, **c** 62.5, **d** 88.4 MeV·cm^2/mg and **e** total number of readouts versus LET

as a single occurrence. The y-axis represents the number of occurrences for each pulse duration indicated on the x-axis. The total occurrences are presented in Fig. 4.25e. Several common trends can be observed across all three devices: Firstly, at an effective LET of 20.4 MeV·cm^2/mg, the maximum voltage drop is insufficient to trigger the amplifier, resulting in an absence of a clear distribution plot in Fig. 4.25a. Secondly, a higher effective LET heavy ion leads to an increase in both the most frequent duration and the overall duration distribution range, as well as the total number of occurrences. This behavior is anticipated since a higher effective LET heavy ion can introduce more ionized charge into the circuit, prolonging the recovery time to the original voltage.

Although most duration readouts in Fig. 4.25b–d display a bell-shaped distribution when the effective LET is greater than or equal to 28.9 MeV·cm^2/mg, sporadic high-duration

4.8 SET Experimental Results

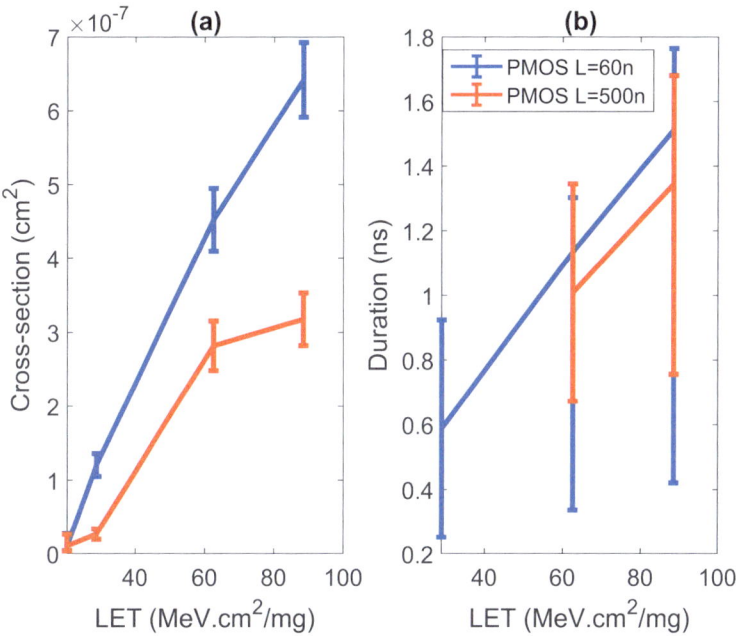

Fig. 4.26 Core PMOS duration measurement results with 1.2 V supply voltage: **a** unit device cross-section, **b** pulse duration range

events occur outside this bell shape. These high-duration events are attributed to double hits occurring within a single measurement period. The influence of channel length on SET duration can be clearly seen from Fig. 4.25c and d. A longer channel results in extended duration due to the increased overall transistor area. When a heavy ion impacts the gate area or even the source area, some charge is collected by the drain through diffusion. This finding aligns with the results from charge measurements.

In the case of applying DNW to the N-type transistor, the total number of occurrences decreases for each effective LET from Fig. 4.25e. Furthermore, both the most frequent duration and the distribution range diminish, which can be clearly notice from 4.25 c and d. This can be explained by two points discussed earlier in the charge measurement results section: the DNW collects a portion of the ionization charge via the local p-well/DNW junction, isolating the charge beneath the DNW and preventing it from being collected [142, 143].

The results of pulse duration measurements for the core PMOS victim devices are illustrated in Fig. 4.26. The unit device cross-section is calculated using the Eq. (4.12) and is displayed in Fig. 4.26a. When the measurement results demonstrate a bell-shaped distribution, the most frequent pulse duration (represented as a square in the figure) along with the upper and lower boundaries (indicated by horizontal bars) are plotted in Fig. 4.26b. Similar trends in duration are observed as the effective LET and channel length vary. It is impor-

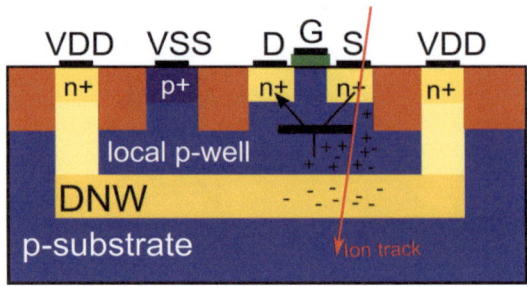

Fig. 4.27 Local p-well potential raising caused by SET

tant to note that the unit device cross-section results obtained from duration measurements are of the same order of magnitude but slightly lower than those from ionization charge measurements shown in Fig. 4.22a. This discrepancy can be attributed to the fact that in the duration measurement circuit, the victim devices are extensively distributed across M blocks ($M > 100$), causing the boundary regions of the victim devices in each block to contribute to a larger sensitive area.

$$CS_{unit,WM} = \frac{Occurrences}{Fluence_{eff} \cdot No.Devices} \quad (4.12)$$

4.8.3.1 IO Devices' Measurement Results

The results of SET pulse duration measurements for IO devices are presented in Fig. 4.28. From the pulse duration range shown in Fig. 4.28b, a similar trend is observed, albeit with lower pulse durations compared to the measurements of core devices. The IO devices operate at a higher supply voltage, resulting in a greater recovery current compared to core devices of identical dimensions.

However, the unit device cross-section results displayed in Fig. 4.28a do not show consistency when compared to the unit device cross-section results obtained from charge measurements. When the effective LET is greater than or equal to $28.9\,\text{MeV} \cdot \text{cm}^2/\text{mg}$, the DNW devices exhibit a larger unit device cross-section than non-DNW devices. This suggests that the presence of DNW enhances the sensitivity of victim devices during pulse duration measurements in high effective LET scenarios.

A similar degradation in SET performance attributed to DNW has also been reported in [145]. This phenomenon may be explained by the potential rise in the p-well resulting from the injection of electrons from the source into the p-well [146]. When an ion strike occurs on the DNW NMOS transistor, electrons and holes are generated through the ionization of the local p-well. The deposited electrons drift into the n-well, while the holes remain in the p-well. The accumulation of holes raises the potential of the local p-well [147].

As illustrated in Fig. 4.27, if the parasitic capacitance of the local p-well is small, this potential rise can eventually lead to forward biasing of the source-p-well, resulting in the injection of electrons into the p-well, which may activate the source-bulk-drain parasitic

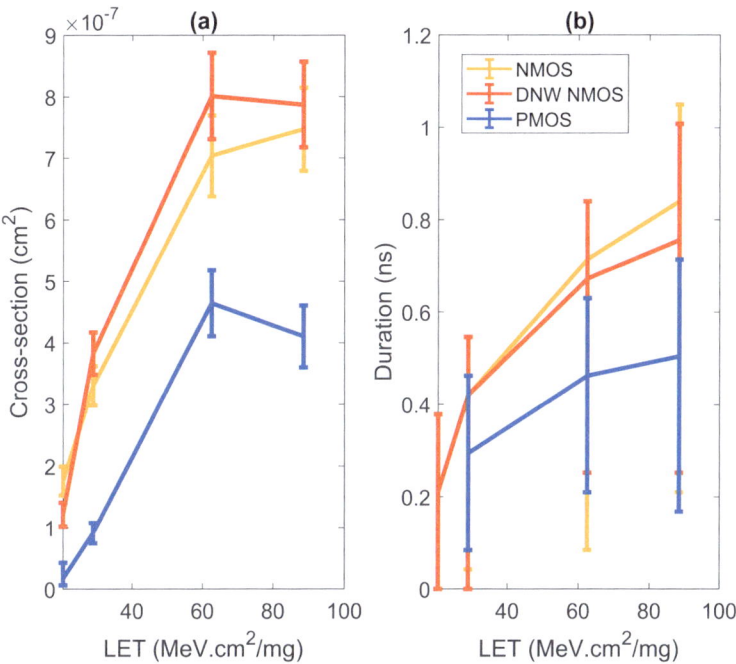

Fig. 4.28 IO NMOS, DNW NMOS and PMOS duration measurement results with L = 500 nm and 3.3 V supply voltage: **a** unit device cross-section, **b** pulse duration range

bipolar transistor [148]. However, most injected electrons are collected by the DNW due to the negligible electric field between the drain and p-well. Nonetheless, some electrons may drift toward the drain, which extends the pulse duration [146].

In the charge measurement circuit, the measurement block includes N victim devices that share one large local p-well. In contrast, for the duration measurement, only N/M ($M > 100$) devices share a smaller local p-well. The p-wells in the charge measurement circuits exhibit a lesser increase in potential when an SET occurs. Consequently, the DNW primarily offers the advantage of reducing charge collection during charge measurements.

4.8.3.2 Measurement Results with Different Supply Voltage

During the SET duration measurement, various supply voltages were applied. The results indicate no significant differences in SET duration when applying supply voltages of 1.08, 1.2, and 1.32 V to the core victim devices. Consequently, only the SET duration of the IO NMOS for both IO NMOS and IO DNW NMOS are presented in Fig. 4.29.

Fig. 4.29 IO NMOS duration measurement results at different supply voltage

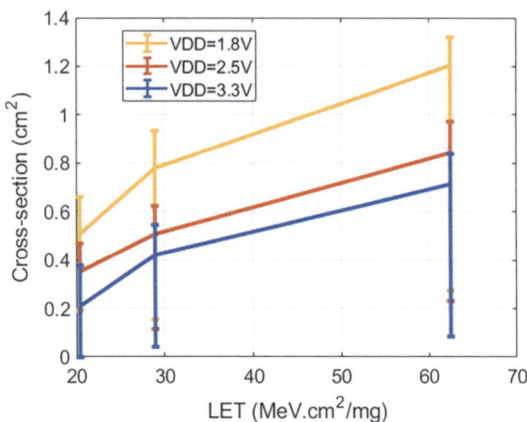

From Fig. 4.29, it is evident that both the most frequent duration value and the distribution range decrease when a higher supply voltage is applied. While a higher supply voltage leads to a thicker reverse junction depletion region and results in more charge being collected, it also increases the recovery current. Consequently, this higher recovery current reduces the pulse duration at elevated supply voltages. A similar trend can be observed in the results for IO DNW NMOS and IO PMOS devices.

4.9 Conclusion

This chapter begins by examining the advantages and disadvantages of advanced CMOS technology concerning radiation effects. Following this, the analysis of 65 nm CMOS technology is presented as a promising candidate for space project development. A test chip utilizing 65 nm technology was developed to characterize the SET ionization charge and pulse duration, and it was tested under a heavy ion beam. The results obtained from the heavy ion tests provide valuable insights for creating an accurate SET current model for radiation-hardened ICs.

In this chip, eight representative core and IO transistors were implemented as victim devices. The study investigates the effects of the deep N-Well structure, channel length, and supply voltages on performance. Measurement results indicate that PMOS transistors exhibit lower charge collection efficiency than NMOS transistors. Furthermore, DNW devices demonstrate a reduced total ionization charge compared to non-DNW devices.

However, in pulse duration measurements, it was observed that the DNW only shortens the pulse duration at low LET values. At higher LET levels, the DNW can actually lengthen the pulse duration due to forward biasing in the local p-well. Additionally, both the channel and source regions of the devices should be considered sensitive areas due to charge contributions through diffusion.

4.9 Conclusion

At elevated supply voltages, an increased amount of charge can be collected; however, this also leads to a rise in the recovery current, consequently shortening the pulse duration. Utilizing the data obtained from the radiation test campaign, the SET current model for 65 nm technology can be updated and applied in the design of subsequent radiation-hardened ADCs.

ADC Fundamentals and Design Tradeoffs

Abstract

This chapter first describes the fundamentals and principles of the analog-to-digital converters. Several conventional ADC structures are analyzed to show their advantages and disadvantages for different requirements and applications. Then, the trade-offs between the conventional ADC and the radiation-hardened ADC are discussed. Four-factor design trade-offs for radiation-tolerant ADC between power, speed, accuracy, and radiation tolerance are concluded. These trade-offs provide fundamental guidelines for the following ADC architecture selection and design.

5.1 Introduction

At the beginning of the twentieth century, research into the conversion of analog signals into digital signals became increasingly important in the field of communication and data transmission. The first application of analog-to-digital conversion was the transmission of telegraph signals [149]. In 1954, the first commercial ADC (11-bit 50 kS/s) in vacuum tubes was introduced by Epsco [150]. After almost 70 years of development, ADCs have become indispensable components that convert analog signals from the real world, such as sound, temperature or voltage, into digital formats that can be processed, stored and transmitted by digital systems. This chapter first introduces the fundamental principle of analog-to-digital conversion. The most commonly used ADC structures are then presented. Then, the noise and non-linearity sources, which degrade the ADC performance, are introduced. Finally, the trade-offs in the development of conventional ADCs and radiation-hardened ADCs are discussed.

5.2 ADC Principle

In analog-to-digital conversion, the analog signal is discretized into digital codes that can be processed by modern digital circuits and systems. The analog signal stands for the time and amplitude-continuous signal in the real world. This conversion can be divided into sampling and quantization, which are shown in Fig. 5.1. The sampler in the ADC receives the input signal and samples it with a fixed sampling period T_s. Another widely used way of expressing the sampling time is using sampling frequency f_s, which is equal to $1/T_s$. The sampler updates its output to the value of the input signal only at the time of nT_s ($n = 1, 2, 3...$). The sampler retains the same value until the next sampling time. In other words, the sampler first samples the input value and then holds it. After sampling, the sampled signal is discrete in the time domain, but its amplitude is still continuous (yellow curve). The continuous amplitude is then approximated by the quantizer with a limited number of steps (blue curve). The total number of output levels of the quantizer is related to the resolution of the ADC. If an ADC has a resolution of N bits, then there are 2^N output levels and the maximum code is $2^N - 1$. The maximum valid input of the ADC is called the full-scale input V_{FS}. If the input is higher than V_{FS}, the ADC is saturated and the output remains $2^N - 1$. The difference in amplitude between two adjacent transition stages is usually referred to as the quantization step (q_s). Note that the difference between the input signal and the quantized signal (yellow and blue curve) is called the quantization error ε_q.

5.2.1 Sampling

As explained in the previous section, the sampler converts the continuous-time signal into the discrete star case signal. The sampling process can be expressed by the following equation:

$$V_s(t) = V_{in}(nT_s) = \sum_{n=0}^{\infty} V_{in}(t)\delta(t - nT_s) \quad (5.1)$$

where $V_{in}(t)$ and $V_s(t)$ are the input signal and the sampled signal, respectively. $\delta(t)$ is the Dirac function, whose integral is equal to one at the integration instant and zero elsewhere [3]. If we perform the Discrete Fourier Transform (DFT) to Eq. (5.1), the sampled signal in frequency domain $V_s(f)$ can be written as follows:

$$V_s(f) = \frac{1}{T_s} \sum_{n=-\infty}^{\infty} V(f - nf_s) \quad (5.2)$$

From the observation of Eq. (5.2), the original signal spectrum is duplicated infinite times, and these replicas are centered at multiples of the sampling frequency f_s. The single-sided sampled spectrum is shown in Fig. 5.2. It is important that f_s is higher than two times f_{in}. In other words, **f_{in} should be less than half of f_s. This is also known as the Nyquist**

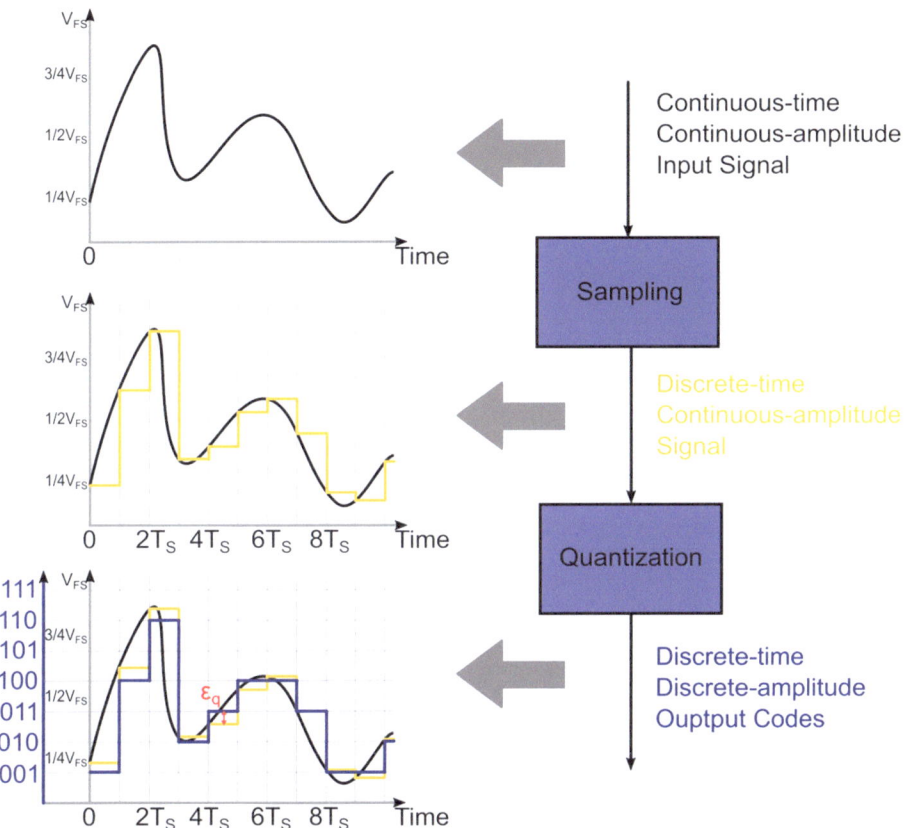

Fig. 5.1 Ideal analog-to-digital conversion principle

criterion [151]. If the Nyquist criterion is not met as shown in Fig. 5.2b, the replica is folded back into the Nyquist zone ($0 \sim \frac{f_s}{2}$) of the spectrum and the aliasing effect occurs. The aliasing effect leads to a loss of information, which means the original signal cannot be reconstructed from the sampled signal.

A signal in the real world contains frequencies up to infinity. Nevertheless, we are only interested in part of the frequency range. For example, the human voice is between 85 and 255 Hz [152], and the voice recorder only needs to process the signal 300 Hz. Therefore, an ADC with a sampling rate of 600 S/s is enough to cover the human voice recording. However, the unwanted high-frequency noise component in the voice signal will cause an aliasing effect and pollute the voice signal after sampling. The most often-used method to avoid the aliasing effect is to connect an anti-aliasing filter in front of the sampler (or ADC). A low-pass filter constructs the anti-aliasing filter and can be active or passive, filtering out the frequency component out of the band of interest. However, to have a steep cutting frequency at $\frac{f_s}{2}$, a multi-order low pass filter is needed, which brings more challenges to the

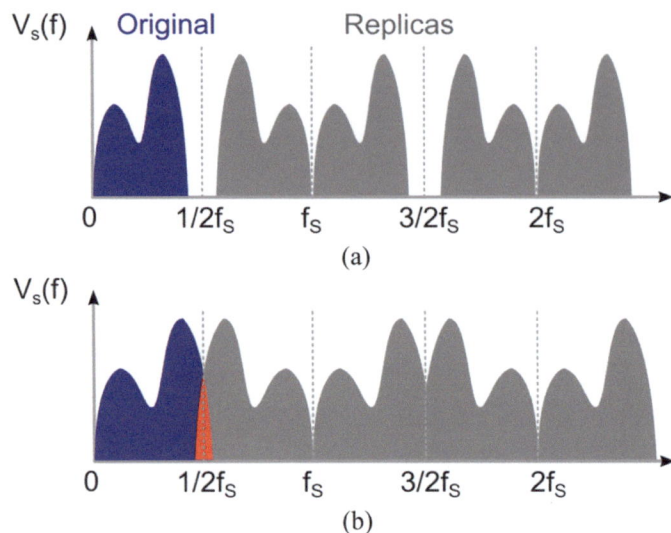

Fig. 5.2 Single side spectrum of the sampled signal: **a** the case of $f_{in} < f_s$, **b** the case of $f_{in} > f_s$

filter design. An alternative method is to increase the sampling frequency of the sampler. If the $\frac{f_s}{2}$ is much higher than the signal frequency, it will lower the requirement of the filter design. However, higher sampling frequency may cause more complexity and less efficiency for the later quantizer design.

5.2.2 Quantization

The sampled continuous amplitude signal is next converted to digital codes by an N-bit quantizer. An N-bit quantizer has 2^N discrete levels, which are used to compare with the analog input. The amplitude of the sampled signal is rounded to the closest discrete level, and the quantizer exports N-digit output codes. The quantization step, which is the amplitude difference between the two adjacent levels, is expressed in $q_s = \frac{V_{FS}}{2^N}$. In some cases, the LSB can also refer to the quantization step. When the input amplitude linearly increases from 0 to V_{FS}, the quantizer will produce a stair shape output digits code. Figure 5.3 describes the ideal transfer characteristic of a 3-bit quantizer. The quantization error ε_q is always between $-\frac{q_s}{2}$ and $\frac{q_s}{2}$ and the digital outputs can be expressed in:

$$V_s(t) = \varepsilon_q(t) + q_s \sum_{i=0}^{N-1} 2^i D_i(t) \quad (5.3)$$

where $D_i(t)$ ($i = 0, 1, ..., N-1$) are the digital outputs of the quantizer at sampling time t. If the input signal frequency does not correlate with the sampling frequency, the error

5.2 ADC Principle

Fig. 5.3 Quantization error: difference between the analog input and digital outputs

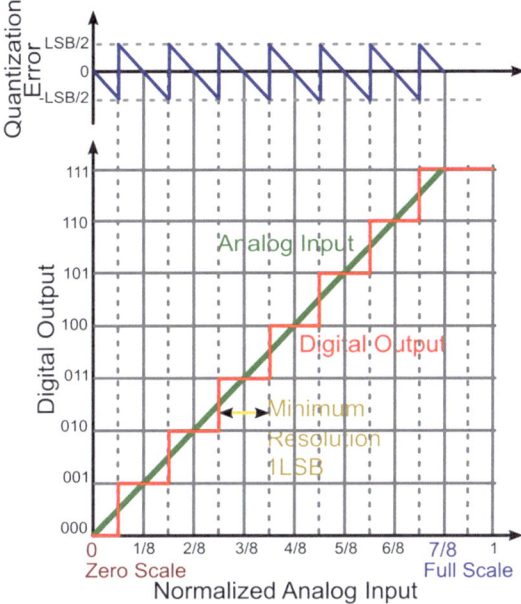

Fig. 5.4 Quantization error probability density

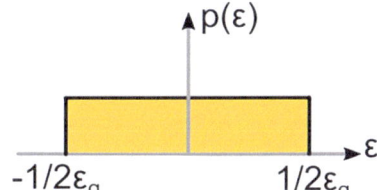

probability should be uniformly spread between $-\frac{q_s}{2}$ and $\frac{q_s}{2}$ (shown in Fig. 5.4). Because of this uniform probability, if a long period of quantization is considered, all errors within $-\frac{q_s}{2}$ and $\frac{q_s}{2}$ will appear the same number of times. The mean squared value of ε_q can be calculated:

$$E(\varepsilon_q^2) = \frac{1}{q_s} \int_{-q_s/2}^{q_s/2} \varepsilon_q^2 \, d\varepsilon_q = \frac{q_s^2}{12} \quad (5.4)$$

The root-mean-square value of the quantization noise can also be calculated:

$$\varepsilon_{q,rms} = \frac{q_s}{2\sqrt{3}} \quad (5.5)$$

If a full-scale input sine wave is applied, whose peak-to-peak amplitude A_{pp} equals $2^N q_s$. The root-mean-square amplitude of the sine wave is:

$$A_{rms} = \frac{2^N q_s}{2\sqrt{2}} \qquad (5.6)$$

The signal-to-quantization-noise ratio (SQNR) corresponding to the ideal N-bit quantizer can be calculated based on Eqs. (5.5) and (5.6):

$$SQNR = \frac{A_{rms}}{\varepsilon_{q,rms}} = 2^N \sqrt{1.5} = 6.02\,\text{N} + 1.76\,\text{dB} \qquad (5.7)$$

Equation (5.7) indicates that for an ideal N-bit quantizer, the maximum SNR equals SQNR. Any non-linearity and the noise in the Nyquist ADC will cause the actual SNR to be lower than the SQNR. Besides, every extra bit added to the quantizer will result in a 6 dB increase in the SQNR.

5.3 Errors in a Non-ideal ADC

The SQNR in Eq. (5.7) is the ideal ratio for an N-bit Nyquist ADC without any error sources, which defines the maximum but not the applicable value of SNR. To make the SNR of a real ADC close to the theoretical limitation, several error sources must be considered during the design. These error sources can be roughly divided into noise errors and non-linearity errors. Noise refers to unwanted random variations or disturbances superimposed on the output digital signal. In ADCs, noise can come from various sources like thermal noise, clock jitter, switching noise, or external interference. Non-linearity refers to deviations from an ideal linear response in the ADC's transfer function. It can manifest as Differential Non-Linearity (DNL) or Integral Non-Linearity (INL). The most obvious difference between noise and non-linearity error is that the noise is random and does not relate to the input signal, while non-linearity error is the opposite.

5.3.1 Noise

- **Thermal Noise**
  Thermal noise, also known as Johnson-Nyquist noise, is a type of random electrical noise that arises due to the thermal agitation of charge carriers (such as electrons) within conductors or resistors. Thermal noise exists in all electronic devices and, thus, also in ADCs. The noise contribution comes from both the sampler and quantizer.
  In the sampler, which is a sample and hold circuit in a Nyquist ADC shown in Fig. 5.5, the thermal noise comes from the on-resistance $R_{S,ON}$ of the sampling switch during sampling [120]. The sampling switch generates white noise with a flat spectral density, which can be expressed in the following:

5.3 Errors in a Non-ideal ADC

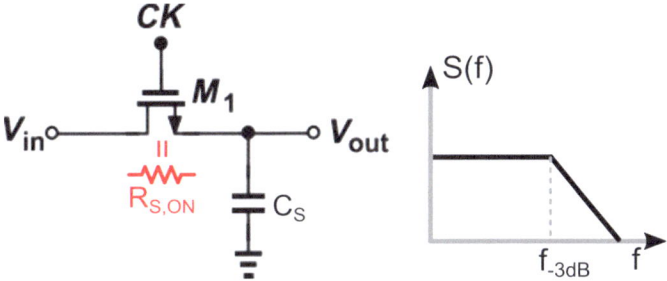

Fig. 5.5 Switch capacitor sample and hold circuit and its cut-off frequency

$$S_{R_{S,ON}}(f) = 4kTR_{S,ON} \tag{5.8}$$

where k is the Boltzmann constant, which equals to $1.38 \cdot 10^{-23}$ J/K, and T is the absolute temperature. The on-resistance of the switch and sampling capacitor C_S form an RC network, which is a first-order low-pass filter. Therefore, the cut-off frequency of such a network is:

$$f_{-3\,\mathrm{dB}} = \frac{1}{2\pi R_{S,ON} C_S} \tag{5.9}$$

Then, the sampler noise power can be calculated by integrating the noise density over the noise bandwidth.

$$\overline{v_{n,samp}^2} = \int_0^\infty \frac{S_{R_{S,ON}}(f)}{(2\pi R_{S,ON} C_S)^2 + 1} df = \frac{kT}{C_S} \tag{5.10}$$

It can be seen that the noise power only relates to the sampling capacitor size. Therefore, a large sampling capacitor can reduce the sampling thermal noise power but also create a large load on the front buffer. Besides, large capacitance also slows the operating frequency of the ADC.

As for the quantizer, the thermal noise can cause comparison errors and then produce error output codes. A typical 1-bit quantizer is a comparator consisting of a pre-amplifier and a series dynamic latch. Because of the pre-amplifier gain, the input-referred noise from the dynamic latch is much reduced [153]. As a consequence, the noise is dominated by the pre-amplifier, and the input-referred noise power is

$$\overline{v_{n,cmp}^2} \approx \frac{kT}{A_{CL,pre-amp}} \tag{5.11}$$

Fig. 5.6 Clock jitter causes sampling error

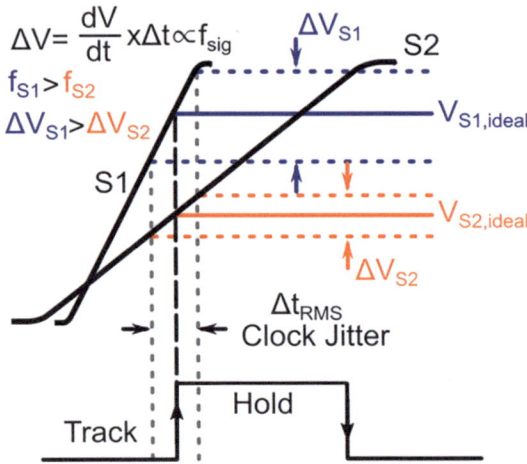

where A is the gain of the pre-amplifier and $C_{L,pre-amp}$ is the capacitive load at the pre-amplifier output. The result indicates that a high pre-amplifier gain can effectively reduce the comparator noise.

- **Clock Jitter Noise**

After the transition from the sampling phase to the hold phase, the input is frozen at the sampling capacitor. This signal freezing behavior of the sample and hold circuit is normally controlled by the rising or falling edge of the sampling clock signal, which is ideal to be $1/f_s$. However, due to the thermal noise of the clock buffer or the non-ideal clock source, the sampling period is not perfectly equal to $1/f_s$ but with a time variation of dt. This time variation then translates into a sampled voltage error dv, whose value largely depends on the input voltage slope. In other words, the dv relates to the input frequency f_{in} as shown in Fig. 5.6. Because of the random sampling error, the maximum SNR is limited by the sampling clock jitter and can be expressed in [154]:

$$SNR_{max} = 20\log(2\pi f_{in}\sigma_{jit}) \qquad (5.12)$$

where σ_{jit} is the RMS value of the clock jitter. Figure 5.7 presents the RMS clock jitter requirements for different f_{in}. It can be seen that for a high input frequency and high wanted output SNR, the ADC sampler asks for a low clock jitter.

5.3.2 Non-linearity

Non-linearity refers to deviations from an ideal linear response in the ADC's transfer function. It can manifest as DNL or INL. An example of INL and DNL characteristics of a 2-bit

5.3 Errors in a Non-ideal ADC

Fig. 5.7 ADC RMS clock jitter requirement

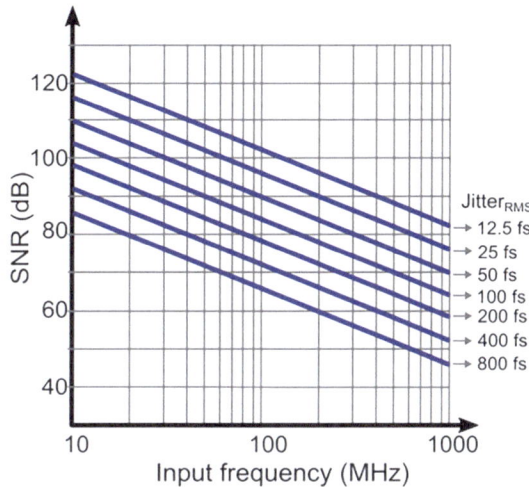

Fig. 5.8 INL/DNL characteristic in a 2-bit ADC

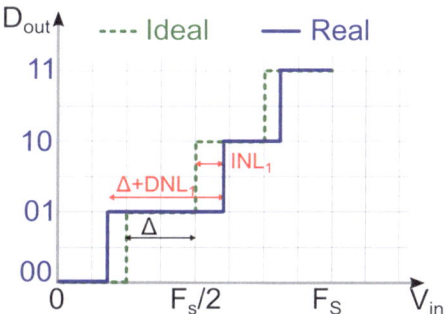

ADC is shown in Fig. 5.8. The DNL error is the difference between the actual step size and the ideal quantization step (q_s) of the ADC, which can be expressed in

$$DNL_i = \frac{(V_{i+1} - V_i) - q_s}{q_s}, i = 0...2^N - 2 \quad (5.13)$$

By sweeping the input DC voltage and measuring the transition voltage V_i, a DNL versus code curve can be plotted. For the real ADC, DNL_i is larger or equal to -1. If a $DNL_i = -1$ is found, which means $V_{i+1} = V_i$, indicates that there is a missing code in the ADC transfer function.

INL measures the deviation of an ADC actual transfer function from an ideal straight line across its entire input range, quantifying the cumulative non-linearity. Because INL describes the overall deviation, it is also the accumulative result of the DNL. The INL can be expressed in

$$INL_j = \frac{(V_{j,real} - V_{j,ideal})}{q_s} = \sum_{i=0}^{j-1} DNL_i, \ j = 0...2^N - 2 \quad (5.14)$$

INL and DNL describe the static error of the ADC output codes. Because the errors are related to the output codes, or in other words, the static error also relates to the input signal. Therefore, the INL/DNL error will cause harmonics in the ADC output spectrum in the dynamic test, which causes SFDR and ENOB drops. The relation between the INL, ENOB and SFDR was estimated in [155] and can be expressed in the following equation:

$$ENOB = \sqrt{1.52}^{[n - \frac{\log(1+3|INL|^2)}{2\log 2}]} \quad (5.15)$$

$$SFDR = 20\log(|INL|2^{-N} + 2^{-1.5N}) \quad (5.16)$$

It must be noted that the equation above is an ideal estimation of the dynamic performance from the static error. In this estimation, the following conditions are assumed:

- Only static error exists in the ADC. All analog and digital blocks are ideal and have no gain or bandwidth limitation.
- The ADC is fully differential, and even order distortion is perfectly rejected.
- The harmonic at the ADC output is mainly contributed by third and fifth-order harmonics. Higher-order harmonics are not considered.

1000 Monte Carlo simulations were performed for a 10-bit ADC to verify the ENOB and SFDR drop due to INL. The results, which are shown in Fig. 5.9, indicate that Eqs. (5.15) and (5.16) can be used as a rule of thumb to estimate the ADC dynamic performance from INL/DNL. In an actual ADC, other effects, such as hysteresis settling error, can also introduce harmonics in the ADC spectrum but are not reflected in the static evaluation (INL/DNL error).

One of the main sources of non-linearity is the component mismatch effect, which occurs in resistors, capacitors and transistors and can lead to nonlinear behavior. For example, in a resistor ladder network used in Flash ADCs, resistor mismatch causes nonuniform reference voltage scaling, resulting in non-linearity. Inaccuracies in the timing of ADC operation, such as clock skew in time-interleaved ADCs, can lead to nonuniform sampling intervals or switching times in different channels, resulting in harmonics in the output. Variations in the voltage/current reference in ADCs also lead to non-linearities in the outputs.

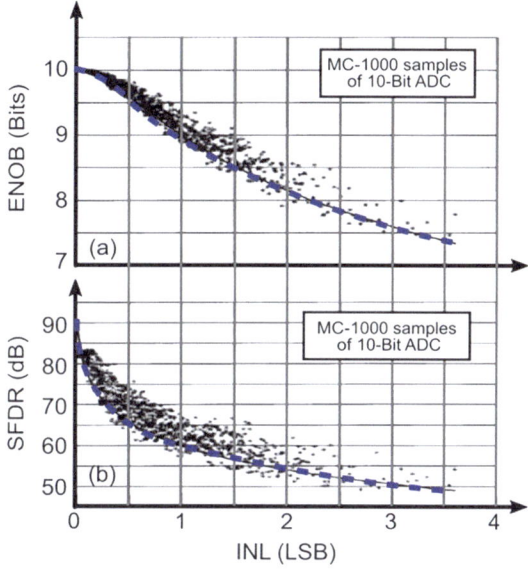

Fig. 5.9 INL/DNL error cause ADC dynamic performance drop: **a** ENOB drop, **b** SFDR drop

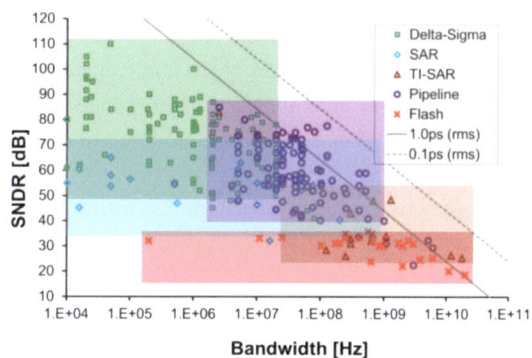

Fig. 5.10 Resolution versus sampling rates in different ADC architectures (*Source* Murmann [13])

5.4 ADC Architectures

In general, ADCs can be divided into two categories, which are shown in Fig. 5.10: Nyquist ADCs and Oversampling ADCs. Nyquist ADCs, such as Flash ADCs, Pipeline ADCs and SAR ADCs, are based on the Nyquist-Shannon sampling theorem, which states that the sampling rate must be at least twice the maximum frequency of the input signal in order to reconstruct the signal without aliasing [156]. Oversampling ADCs, such as delta-sigma ADCs, sample the input signal at a rate that is well above the Nyquist frequency. By sampling at a much higher rate, oversampling ADCs capture more samples than necessary, providing higher resolution.

5.4.1 Flash ADCs

If the application requires a fast sampling rate (gigasamples per second), the Flash ADC may be a suitable candidate. The Flash ADC is one of the simplest structures, as it compares the input signal directly with all references, without a sample and hold block [157]. In a resistor-based N-bit Flash ADC, 2^{N-1} comparators are connected in parallel and 2^N resistors are connected in series. As the resolution of a flash ADC increases, the number of comparators required increases exponentially, leading to greater circuit complexity and higher power consumption. In addition, the matching between comparators and resistors in high-resolution flash ADCs limits the static and dynamic performance.

5.4.2 Pipeline ADCs

Pipeline ADCs process analog input signals by dividing the conversion process into successive stages [158]. Figure 5.11 shows a typical pipeline ADC structure with N stages, where each stage processes M bits. The first stage samples the analog input signal during the sampling process and then transforms a portion of the signal into digital codes. The analog signal is then recreated by the stage's sub-DAC and the residue is generated by subtracting the DAC output from the sampled input signal. The residue is then amplified with a gain of 2^M and fed to the next pipeline stage. The procedure is repeated in the next stage. Pipeline ADCs achieve a high sampling rate, but are slower than flash ADCs due to the sequential conversion. Pipeline ADCs have a better resolution as they require fewer components. However, pipeline structures have a higher design complexity than other ADC architectures due

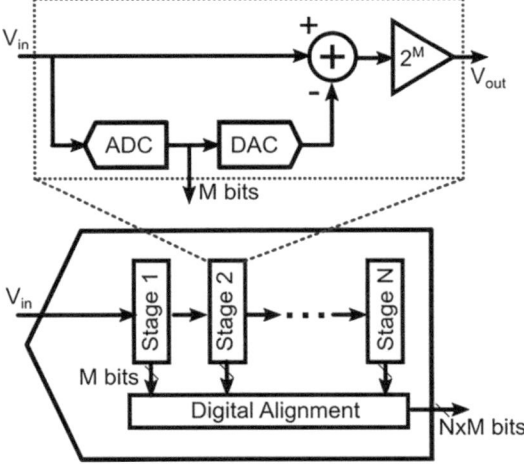

Fig. 5.11 Pipeline ADC structure

to the sub-DACs and stage amplifiers. Latency is another issue with pipelined structures as each sample requires N sample periods to generate digital codes.

5.4.3 SAR ADCs

The SAR (Successive Approximation Register) architecture, shown in Fig. 5.12, has a relatively simple architecture consisting of comparators, a DAC and a logic control block [159]. In principle, SAR ADCs start by sampling the analog input voltage and then iteratively determine each digital output bit by comparing the generated DAC reference voltage to the sampled input. Starting with the MSB, it assumes bit values and adjusts them based on the output of the comparator. This iterative process continues for each bit, successively refining the digital code until the closest approximation to the analog input is achieved. Due to its simple and almost exclusively digital structure, the SAR structure offers several advantages. SAR ADCs can achieve moderate resolution and speed with high power efficiency. They have low design complexity and are process scalable. However, SAR ADCs are unsuitable for high-speed applications since N conversion cycles are required for N-bit resolution. Besides, with high-resolution requirements, a large number of components in the DAC can cause a serious matching problem.

5.4.4 Sigma Delta ($\Sigma\Delta$) ADCs

The Sigma Delta ($\Sigma\Delta$) ADCs take a fundamentally different approach to the Nyquist ADCs mentioned above. Figure 5.13 shows a first-order $\Sigma\Delta$ ADC structure consisting of a $\Sigma\Delta$ modulator and a digital output filter/decimator. $\Sigma\Delta$ ADCs oversample the input signal at a very high frequency, significantly above the Nyquist frequency [161]. They employ noise-shaping techniques, particularly high-order delta modulation, to push quantization noise out of the frequency band of interest and effectively shift it to higher frequencies. The modulator converts the analog input into a digital bit stream, while the digital filter converts the bit

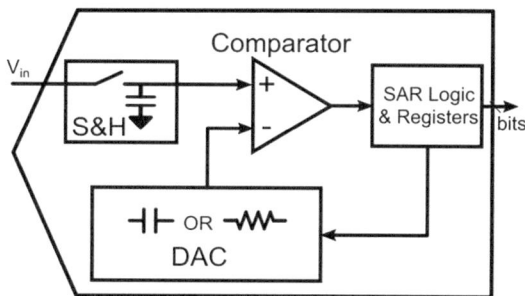

Fig. 5.12 Successive approximation register ADC structure

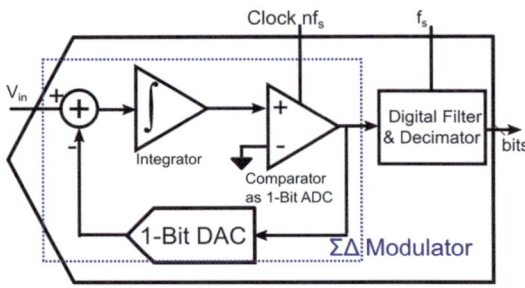

Fig. 5.13 First-order delta-sigma ADC structure (*Source* Kester [160])

stream into a data word representing the magnitude of the analog input. This bit stream is then digitally filtered and decimated, resulting in a binary format. Since $\Sigma\Delta$ ADCs exchange resolution with an oversampling ratio, they are very well suited for applications with low input signal bandwidth but requiring high output resolution.

5.4.5 Hybrid ADCs

Hybrid ADCs combine the features of multiple ADC architectures in a single device to leverage the strengths of each architecture while mitigating their limitations. Pipelined-SAR ADCs combine the speed of the pipeline ADC with the resolution and linearity advantages of a SAR ADC [162]. This structure consists of multiple stages, and each stage incorporates SAR-based conversion techniques, which allows for parallel processing across stages to reach a higher sampling frequency. Noise-shaping SAR ADCs merge the high power efficiency of SAR ADCs and high SNR from $\Sigma\Delta$ ADCs [163].

5.5 ADC Design Tradeoff

The speed-power-accuracy tradeoff is a fundamental consideration in ADC design. Speed, which refers to the conversion rate, is the amount of time it takes to convert an analog signal to a digital format. The power is calculated from the total current of the supply voltage during the conversion. Accuracy indicates how accurately the digital output codes can reproduce the analog input signal. In practice, covering all three factors in the tradeoff is difficult. Pushing one factor better will cause at least one of the other two factors to degrade.

A higher conversion speed requires a high gain bandwidth, which is limited by the dominant pole of the circuit system. Therefore, either (both) increasing the bias current or (and) decreasing the node capacitance can help to increase the conversion speed. However, a higher current directly increases the total power consumption of the ADC. Lower capacitance demands for smaller component dimensions, which results in poorer device matching and lower accuracy. Figure 5.14a, b show the accuracy and power consumption as a function of

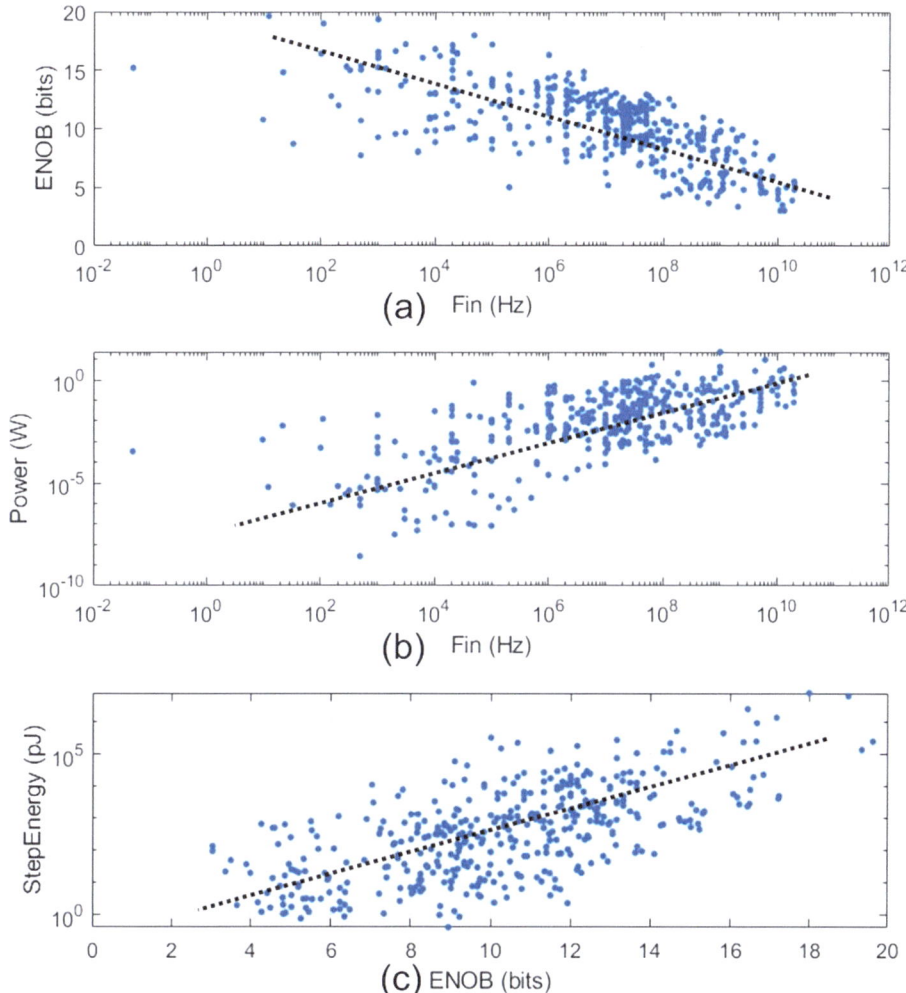

Fig. 5.14 ADC speed-power-accuracy tradeoff from ADC survey: **a** speed versus accuracy, **b** speed versus power, **c** accuracy verse energy (*Source* Murmann [13])

the input signal frequency. A higher input frequency or sampling frequency leads to lower accuracy and higher power consumption.

Without taking the speed into account, the conversion accuracy is also related to the conversion energy per sample (P/f_s). Higher conversion accuracy requires low mismatch caused by large dimensions and high linearity of the active circuits. Figure 5.14c demonstrates the accuracy as a function of ADC sampling energy in previously published work.

The trade-off between speed, power and accuracy leads to a variety of ADC architectures optimized for different applications, from high-speed applications to high-resolution and

Fig. 5.15 Spider diagram of ADC architectures, design tradeoffs

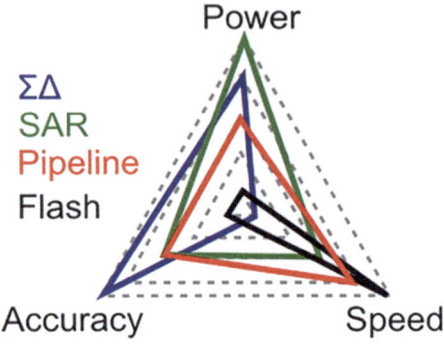

low-power applications [164]. Figure 5.15 shows the speed-power-accuracy performance of the conventional ADC structures mentioned in the previous section.

5.6 Radiation-Hardened ADC Design Tradeoffs

Space radiation mainly comprises high-energy particles such as protons, electrons, and heavy ions originating from the Sun, galactic cosmic rays, and trapped particles in the radiation belts around astronomical objects [27]. When these energetic particles interact with electronic devices such as ADCs, performance degradation or malfunction may occur. The radiation robustness can be exchanged with ADC power, speed, and accuracy or improved by using RHBD techniques. In all cases, this comes at the expense of energy efficiency. Therefore, the design tradeoffs of radiation-hardened ADC are analyzed to achieve an energy-efficient ADC design. In recent years, the trade-offs in conventional ADCs have been well explored and balanced by many previous works that achieve excellent performance efficiency. However, the design tradeoffs and strategy of radiation-hardened ADCs are still unclear. For TID effects, the main degradation happens to the parameters of the transistors, like threshold voltage and leakage. Therefore, TID effects can be treated as a special corner of the transistors and analysis. However, for SEEs, the analysis of the tradeoffs can be much more complex and challenging since ADCs are analog/mixed-signal components. Different consequences and effects can happen to different radiation effects.

A S&H circuit, one of the simplest analog/mixed-signal circuits, is used as an example to explain the difficulty of analyzing the radiation response. When it meets the radiation effects, its response strongly relates to the states of the operation, what effect and how severe the effect is. Figure 5.16 shows a simple S&H circuit, which consists of a digital block to generate the clock signal, a transmission gate, and a sampling capacitor, which are an analog RC circuit. The ideal waveform is also shown. The capacitor starts to track the input when the control clock goes high and then freezes the input at the falling edge until the next rising edge. When discussing the radiation effects in such circuits, we need to discuss

5.6 Radiation-Hardened ADC Design Tradeoffs

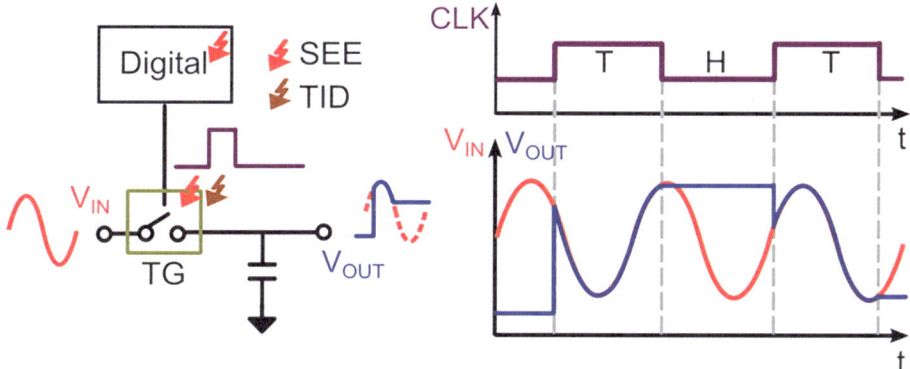

Fig. 5.16 Simple Sample and Hold (S&H) circuit in an ADC with ideal waveform

Fig. 5.17 Voltage on sampling capacitor draft at hold phase due to TID effects

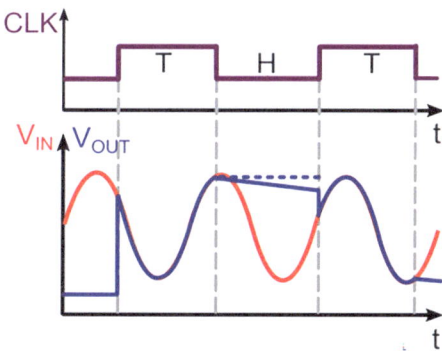

both TID effects and SEEs. TID effects are straightforward: TID effects mainly affect the S&H in the hold phase. The stored charge may leak away faster due to leakage increase and threshold shift as shown in Fig. 5.17.

However, analysis becomes complex when such circuits experience SEEs. The consequence depends on the location, time, and servility. There are two parts in S&H that may experience SEE errors, which are clock generation (producing the control signal, logic high for tracking (T) and logic low for holding (H)) and transmission gate, which is a switch. When the SEE happens to the clock generation and the ionized charge is sufficient to cause a flip, the error cases at the output clock can be concluded in Fig. 5.18. It can be seen that, in cases (a) and (c), SEEs can cause extra hold and track phases in a single sampling period. Even case (a) may still produce the right output in the original hold phase, extra phases can cause serious issues in ADC digital control since most of the internal operation is triggered by the sampling clock. As for case (b), the track phase is shorter, and the actual sampling point is shifted (the falling edge), which causes sampling error.

For the transmission gate, SEE still causes errors even when the control signal is correct. When SEE happens to the transmission gate at the tracking period, assuming the input buffer

Fig. 5.18 S&H control clock error due to SEEs at different phases

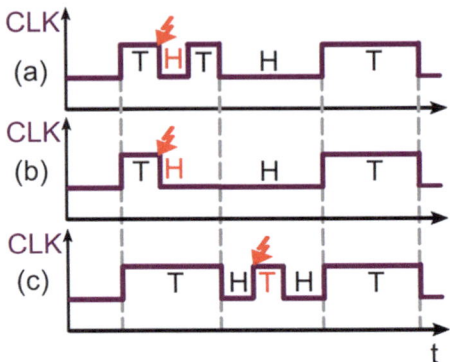

is strong enough, the outputs are not affected by the SEE since the input buffer will "flush away" the ionization charge generated by energy particles as shown in Fig. 5.19a. On the contrary, if the amount of the ionized charge is large enough or the SEE happens at or close to the end of the sampling period, this error charge will be frozen to the capacitor and cause a sampling error, which is shown in Fig. 5.19b. When an SEE happens at the hold phase to the sampling switch, the outputs become untrustworthy since the frozen charge is always changed. From the above example of an S&H, it is hardly to conclude a simple rule to improve the radiation since the radiation error depends on so many parameters. To simplify the analysis, we first divided the radiation effect by time scale: long-term effects and short-term effects, which are TID effects and SEEs, respectively [27, 165]. Then, for SEEs in ADC, we further analyze the tradeoffs in analog and digital blocks.

5.6.1 Total Ionizing Dose Effects

TID effects, which are long-term effects, occur when the MOSFETs are exposed to ionizing radiation over a certain period of time, which can last from months to years. The detailed explanation can be found in Chap. 2. Charge accumulation in the silicon dioxide (SiO_2) and in the field isolation oxide leads to a threshold shift ΔV_{th} and parasitic transistor leakage I_{para} [46].

$$\Delta V_{th} \propto Dose \cdot t_{ox}^2 \qquad (5.17)$$

$$I_{para} \propto Dose \cdot \frac{1}{L} \qquad (5.18)$$

where $Dose$ is a measure of the cumulative ionizing radiation absorbed by the semiconductor material over time. t_{ox} is the oxide thickness and L is the transistor channel length.

The working conditions, or bias conditions, of MOSFETs can be defined as on-state (strong inversion) and off-state (weak inversion). The saturation state of MOSFETs is mostly

5.6 Radiation-Hardened ADC Design Tradeoffs

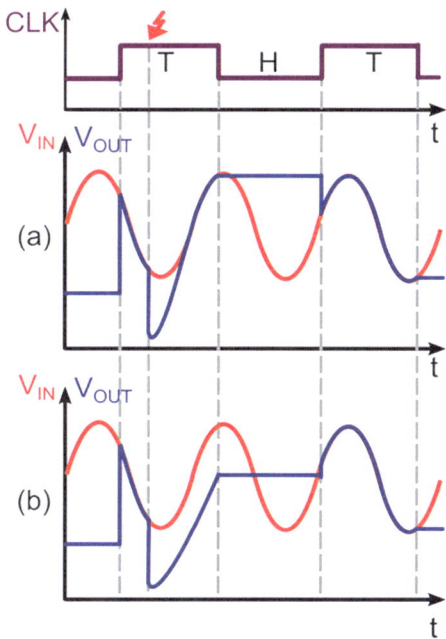

Fig. 5.19 SEE happens at the TG of a S&H circuit

used in analog design. To simplify the analysis, only the strong inversion with saturation state is therefore discussed here. Considering the degradation due to TID effects, the drain-source current can be expressed with the TID degradation in Eqs. (5.19) and (5.20).

$$I_{DS,wi} = I_0 exp(\frac{V_{GS} - (V_{th} + \Delta V_{th})}{nkt/q}) + I_{para} \quad (5.19)$$

$$I_{DS,si} \approx \frac{W\mu C_{ox}}{2L}[V_{GS} - (V_{th} + \Delta V_{th})]^2 + I_{para} \quad (5.20)$$

Both the threshold drift and the parasitic transistor leakage current have a significant impact on the transistor in the off state, as shown in Eq. (5.19). This leads to a drastic increase in the power consumption of the chip, especially in the digital blocks. Therefore, a longer transistor length is preferred to minimize this leakage current. As a result, the digital blocks in the ADC have a lower operating frequency and higher switching current consumption due to the larger dimensions. In the saturation state, threshold shift is the dominant degradation that has a greater impact on analog blocks such as the amplifiers in the ADC. Amplifiers normally require a stable operating point to achieve stable gain and loop stability. However, a threshold shift can move the transistors away from the desired operating point and lead to degradation in linearity, speed and noise. Therefore, transistors with strong inversion and high V_{dsat} ($V_{dsat} = V_{GS} - V_{th}$) are preferred to achieve stable performance in the entire dose range, which is less power efficient as the weak and moderate

Table 5.1 ADC design tradeoffs considering TID effect tolerance

TID-introduced Issues	Solutions	Speed	Power	Accuracy	Design Effort
Leakage increase	Longer transistor channel length	↓	↑		
Threshold shift	Saturation with high overdrive voltage V_{dsat}		↑		↑
1/f noise increase	Saturation with high overdrive voltage V_{dsat}		↑		↑

inversion can provide higher transconductance efficiency. However, thanks to the scaling of the technology, the TID effects are mitigated due to the thinner gate oxide and smaller field isolation oxide [68]. When process nodes drop to 180 nm or below, most circuits can pass 100 krad(Si). Deep submicron technology nodes (90 nm and below) can basically tolerate 300 krad(Si) or even Mrad values [58]. Therefore, the digital blocks are much less affected by TID effects. However, the strongly biased transistor also has a worse degradation when experiencing the same total dose. From the results of 65 nm technology, a total dose of 400 Mrad can cause more than a 5x threshold shift difference between the zero biased and fully active transistors (Vgs = Vds = VDD) [127].

Besides the voltage and current degradation, TID effects can cause the 1/f noise degradation to MOSFETs. The 1/f noise spectra density can be expressed in Eq. (5.21) [166]. In the equation, the normalized noise power factor K increases when the interface and oxide trap are increased after irradiation [167, 168]. Larger overdrive voltage and dimensions (W and L) help to reduce the flicker noises. The above-mentioned tradeoffs considering the TID effects are concluded in the Table 5.1.

$$S_v \propto K \cdot Dose \cdot \frac{(V_{DS})^2}{(V_{GS} - V_{th})^2} \cdot f^{-\alpha} \qquad (5.21)$$

5.6.2 Single Event Effects

Short-term effects, known as SEEs, occur when a high-energy particle interacts with sensitive regions of a microelectronic device. The particle's passage generates free charge carriers through ionization. If these charge carriers are collected by the MOSFETs, a SET voltage

5.6 Radiation-Hardened ADC Design Tradeoffs

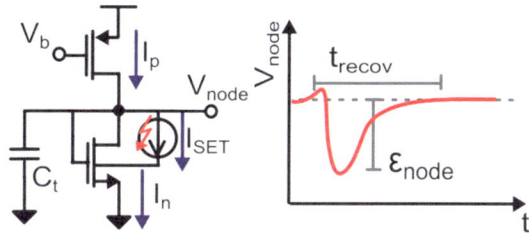

Fig. 5.20 Single event effects in a simple two-transistor model

disturbance is induced [72]. Depending on the circuit's topology, functionality and SET pulse width/depth, SET consequence can be much different and hard to predict in an ADC. The SEE may result in effects such as SEU, SEL, or SEFI, which are detailed in Chap. 2. Due to the uncertainty of the effects, a careful study of the SET effects is required for every concrete ADC design [169]. To make the analysis simpler, the following analysis starts from the basic circuit topology—two-transistor amplifier/inverter, which is widely used in both the digital and analog domains and is shown in Fig. 5.20. In this example, the gate of the NMOS transistor, which is the input, is tied to ground, and the top PMOS transistor is connected to a known bias V_b. When an SEE happens to such a circuit, the output voltage $V_{out}(t)$ of such a circuit can be modeled by incorporating an SET current source, $I_{SET}(t)$, expressed as a double exponential model [170]:

$$V_{out}(t) = V_{out}(t_0) + \int_{t}^{t+t_0} \frac{I_p(t) - I_n(t) - I_{SET}(t)}{C_t} dt \quad (5.22)$$

$$I_{SET}(t) = \frac{Q_{SET}}{\tau_a - \tau_b} \left[e^{-\frac{t-t_0}{\tau_a}} - e^{-\frac{t-t_0}{\tau_b}} \right] \quad (5.23)$$

where C_t represents the total capacitance at the output node, and $I_p(t)$ and $I_n(t)$ denote the current contributions from the PMOS and NMOS transistors, respectively. The variable t_0 indicates the moment when the heavy ion interacts with the node V_{out}. The total ionized charge generated by the heavy ion, Q_{SET}, depends on factors such as the particle energy, manufacturing process, and supply voltage. The parameters τ_a and τ_b correspond to the rising and falling time constants of the SET, determined by the process technology.

To simplify the analysis, it is assumed that the ionized charge is deposited in a very short duration, and the NMOS transistor is turned off due to the voltage drop at V_{out}. Under these conditions, the error voltage amplitude $\varepsilon_{out}(t)$ at the output node can be expressed as:

$$\varepsilon_{out}(t) = V_{out}(t) - V_{out}(t_0)$$
$$\approx -\frac{1}{C_t}(Q_{SET} - I_{p,\max} \cdot t) \quad (5.24)$$

where $I_{p,\max}$ represents the maximum current supplied by the PMOS transistor to compensate for the ionization charge. The recovery time, t_{recov}, is defined as the time required for

V_{out} to return to its pre-SEE value, which is determined by solving $\varepsilon_{out}(t_{recov}) = 0$. If V_{out} is reset by the sampling clock with frequency f_s, the recovery time is constrained by the clock period $T_s = 1/f_s$:

$$t_{recov} = \begin{cases} \dfrac{Q_{SET}}{I_{p,\max}} & \text{if } V_{out} \text{ is not reset by } f_s \\ \max\left[\dfrac{Q_{SET}}{I_{p,\max}}, T_s\right] & \text{if } V_{out} \text{ is reset by } f_s \end{cases} \qquad (5.25)$$

When there is no SEE, the ADC has one signal input, which is V_{in}, and generates the correct output $D_{out}(nT_s)$ as shown in Fig. 5.21a. However, when SEE happens to ADC, the voltage disturbance caused by the SEE propagates through the ADC and results in erroneous output codes. Therefore, all internal circuit nodes can be the input and produce error outputs. ADC in this case is a multiple-input single-output component. The final output error of the ADC is influenced by the specific location where the SEE occurs. The ADC sub-blocks can be categorized into analog and digital sections, each with distinct error mechanisms due to SEEs.

For the analog components as shown in Fig. 5.21b, the ADC output error is determined by the least significant bit (LSB) voltage of the ADC and the gain factor, A_{out}, between the impacted node and the ADC output. This relationship can be expressed as:

$$D_{err,ana}(nT_s) = \frac{\varepsilon_{out}(t)\delta(t - nT_s)A_{\text{out}}}{LSB} \qquad (5.26)$$

where A_{out} represents the functionality and sensitivity of the analog block. It also depends on the specific timing of the disturbance within the conversion cycle. Because of A_{out}, even a minor disturbance can amplify into a significant error in the output, measured in terms of LSB.

Digital blocks (as shown in Fig. 5.21c) are generally less susceptible to SEEs compared to analog blocks, as they operate solely with high and low voltage levels. However, if the maximum value of $\varepsilon_{\text{out}}(t)$ exceeds a certain threshold, the voltage disturbance caused by a heavy ion strike can result in either a transient bit flip in combinational logic or a permanent bit flip in sequential logic. The transient flip manifests as a voltage pulse, commonly referred to as a DSET, while the permanent bit flip is classified as an SEU.

To induce a DSET or SEU, the peak amplitude of $\varepsilon_{\text{out}}(t)$ must exceed the node's flip threshold, $V_{\text{th,flip}}$. These events, DSET or SEU, lead to error codes $\varepsilon_{\text{err,dig}}(nT_s)$ in the final ADC outputs, which can be expressed as:

$$D_{err,dig}(nT_s) = \begin{cases} 0 & \max[\varepsilon_{out}(t)] < V_{th,flip} \\ \varepsilon_{fun}(nT_s) & \max[\varepsilon_{out}(t)] \geq V_{th,flip} \end{cases} \qquad (5.27)$$

The amplitude and duration of $\varepsilon_{\text{err,dig}}(nT_s)$ are influenced by the functionality of the digital blocks within the ADC. In certain blocks, such as the output code shift register, the amplitude of $\varepsilon_{\text{err,dig}}(nT_s)$ can vary from 1 LSB to 2^{N-1} LSBs and typically affects only a single

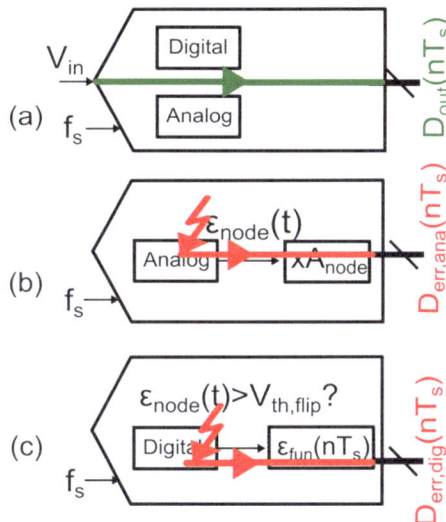

Fig. 5.21 An ADC has **a** no radiation effect, **b** an SEE in the analog block, and **c** an SEE in the digital block

sample. However, in critical blocks like clock generation, $\varepsilon_{\text{err,dig}}(nT_s)$ can persist indefinitely, resulting in a complete functional failure of the ADC.

The SEE tolerance of an ADC can be enhanced through three primary approaches, which are concluded in Table 5.2. The first approach involves mitigating the source of the error, namely the SET error pulse. This can be achieved by increasing the node capacitance, sampling period, or bias current, as described in Eqs. (5.24) and (5.25). However, these modifications typically reduce the ADC's power efficiency and speed. The second approach focuses on improving the analog SEE tolerance. This can be done by increasing the LSB voltage or decreasing the node sensitivity as outlined in Eq. (5.26). While a higher LSB voltage improves tolerance, it requires a higher power supply or results in a lower ADC resolution. Lastly, digital blocks inherently exhibit greater tolerance compared to analog blocks, as $\varepsilon_{\text{out}}(t)$ must exceed $V_{\text{th,flip}}$ to induce a DSET or SEU. Enhancing this tolerance can be achieved by increasing the threshold voltage through added capacitors.

5.6.3 ADC Structure Selection Consideration

It can be inferred from Tables 5.1 and 5.2 that achieving radiation tolerance typically necessitates a low sampling frequency and low-resolution ADC. Additionally, a lower clock frequency is generally preferred because circuit states are influenced by clock transitions. Reducing the clock frequency allows more recovery time for SEE-induced errors before they propagate into the system. Consequently, Nyquist ADCs naturally exhibit a lower likelihood of capturing SEE-induced errors compared to oversampling ADCs when operating at the same output data rate [171].

Table 5.2 ADC design tradeoffs considering SEE tolerance

SEE error Level	Solutions	Speed	Power	Accuracy	Design effort
SEE source	Lager node capacitance	↓	↑		
SEE source	Higher biasing current		↑		
ADC analog	Lager LSB voltage		↑	↓	
ADC analog	Lower analog node gain A_{out}				↑
ADC Analog/Digital	Longer conversion period	↓			
ADC Digital	Higher digital $V_{th,DSET}$	↓	↑		
ADC Digital	Reset by master clock				↑

An effective strategy to mitigate the speed penalty involves employing multiple low-sampling-rate ADCs in a time-interleaved configuration, rather than relying on a single high-sampling-rate Nyquist ADC. A similar approach can enhance ADC resolution by employing a pipelined architecture, where a high-resolution ADC is segmented into multiple low-resolution stages with interstage residue amplifiers (RAs). As a result, the time-interleaved pipelined structure offers superior inherent radiation tolerance compared to single-stage ADCs. However, certain sub-blocks remain vulnerable to radiation effects. For instance, the clock generation and output code alignment circuits continue to operate at the reference clock frequency, and the inclusion of RAs introduces additional analog components, which may increase sensitivity. To address these vulnerabilities, targeted hardening is required for such blocks. With careful design, the performance loss due to hardening is minimal. Details of the hardening strategies for these critical blocks are discussed in subsequent sections.

5.7 Conclusion

This chapter first describes the basic principle of analog-to-digital conversion. Several conventional ADC structures are analyzed to show their advantages and disadvantages for different applications. Next, the design trade-offs between the conventional ADC and the radiation-hardened ADC are discussed. In conclusion, the radiation tolerance of an ADC inevitably leads to a degradation in performance, speed, or accuracy. This trade-off provides fundamental guidelines for the following system-level ADC design.

Radiation Hardened Pipelined-SAR ADC Architectural Modeling and Design Considerations

6

Abstract

This chapter presents the system-level analysis of a pipelined-SAR ADC and determines the optimal bit splitting between the coarse and fine stages of the ADC to achieve optimal power efficiency. Various CDAC switching schemes are also investigated to further minimize the power consumption of the ADC. At ADC resolution higher than 12 bits, mismatch effects limit the ADC linearity and increase the CDAC sampling capacitance, which degrades the sampling frequency. Therefore, a novel mismatch calibration method is proposed to minimize the harmonics caused by mismatch. At the end of this chapter, all sub-blocks with the TID effects and SEEs are analyzed. The TID has a smaller effect on the pipelined-SAR ADC in the 65 nm process thanks to the process scaling. However, for the SEEs, the residual amplifier and some digital circuits, such as the clock generation block, can cause multiple errors or even unrecoverable failure of the ADC outputs. They need to be hardened to increase the overall SEE tolerance.

6.1 Introduction

The previous chapter introduced the basic theory and architectures of ADCs. The trade-offs in ADC design between power, resolution, speed and radiation tolerance were also explored. This chapter presents the system-level design and considerations for the proposed state-of-the-art radiation-tolerant ADC in 65 nm technology. This ADC aimed to achieve a conversion speed of at least 50 MS/s with an ENOB of 11 bits. The total power consumption of the ADC should be less than 50 mW.

6.2 SAR-Assisted Pipeline Structure

Among the various ADC structures, ADCs with SAR are attractive due to their moderate accuracy and speed, simple structure, low power consumption and technological scalability [159, 172, 173]. Above all, almost all digital structures of SAR ADCs can be hardened more easily against radiation effects. However, with an accuracy of more than 12 bits and sampling rates of more than 10 MS/s, the performance of SAR ADCs is limited by the sampling settling time and the matching requirements of the capacitor bank. Each additional bit doubles the number of capacitors and also lowers the sampling rate [173]. Alternatively, the pipelined-SAR ADCs shown in Fig. 6.1, which include a coarse-stage ADC (C-ADC), a residual amplifier (RA) and a fine-stage ADC (F-ADC), solve the capacitor bank problem and increase the accuracy and sampling rate [162, 174]. It lowers the speed and mismatch requirements of the SAR architecture by splitting the conversion process into stages while maintaining accuracy and resolution through the SAR algorithm at each stage. This combination enables high-speed conversions at relatively high resolutions, making it suitable for applications that require both speed and accuracy in analog-to-digital conversion. Furthermore, the pipelined-SAR architecture offers additional advantages for radiation tolerance. First, according to Eq. (5.26), additional SEE tolerance is provided by dividing a high-resolution ADC into two or more low-resolution sub-ADCs. Due to the sequential conversion into coarse and fine stages, all sub-ADCs are relaxed in their operating bandwidth and can tolerate a longer SEE recovery time from Eq. (5.25) as they operate at f_s.

All sub-blocks of a pipelined-SAR ADC are shown in Fig. 6.1. The input signal is first sampled by the CDAC and converted into the first MSBs while the residual voltage is generated. Since the sampled voltage has a direct effect on the convergence accuracy, the sampling capacitor (total capacitance of the CDAC of the coarse stage) must correspond to the total resolution of the ADC, such as the thermal noise requirement. In addition, the mismatch of the CDAC in C-ADC leads to distortions in the residual voltage and to harmonics in the fine stage. Therefore, the mismatch of the CDAC in C-ADC must also fulfill the overall resolution, which can be achieved by implementing the mismatch calibration. Due to the sampling and mismatch requirements, the switching power of the CDAC is one of the main factors affecting the power efficiency of the C-ADC. Therefore, a suitable switching scheme is required to minimize the switching power. In addition to the C-ADC, the RA amplifies the residual voltage, and the amplification ratio is determined by the number of bits in the C-ADC and the input range of the F-ADC. Because of the high amplification ratio, the input offset of the RA can reduce or even saturate the F-ADC, which requires an offset calibration such as auto-zeroing offset cancellation. After the amplified residue voltage has been sampled, the F-ADC processes the remaining bits. Since the F-ADC processes only the last LSBs of the total resolution, the sampling noise and mismatch requirements are much lower than the C-ADC, and the CDAC of the F-ADC can be much smaller to save power. For the comparators in C-ADC and F-ADC, the input-referred noise only needs to meet the requirements for the sub-ADCs. In addition to the two sub-ADCs and RA, there are several

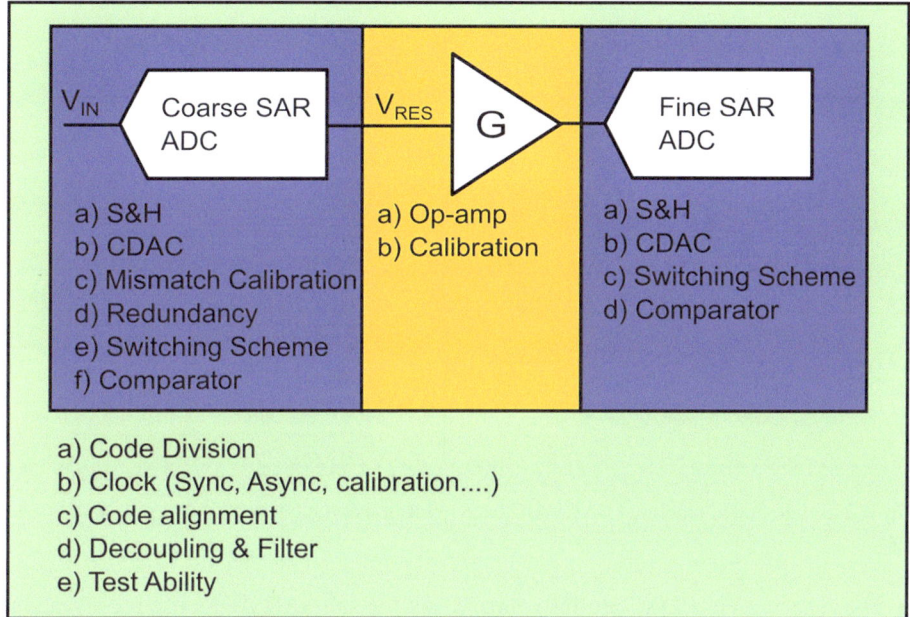

Fig. 6.1 Typical pipelined-SAR ADC structure and its sub-blocks

blocks that serve the pipeline structure: The clock generation block produces the sampling clock and the comparison clock for C-ADC and F-ADC, as well as the amplification control for RA. Code alignment is used for combining the codes of C-ADC and F-ADC. The consideration of radiation in the sub-blocks is covered in the last part of this chapter.

6.3 Bit Division Modeling in SAR-Assisted Pipeline ADC

In an N-bit SAR-assisted pipeline ADC, the bit splitting between the coarse (M bits) and fine stages ($N - M$ bits) can cause a difference in power efficiency and sub-block design complexity. A system-level single-ended pipelined-SAR ADC model (see Fig. 6.2) was created in Matlab to analyze the optimal bit splitting for minimum power consumption, which consists of the power consumption of two sub-ADCs (CDAC switching power, comparator power, logic power) and the residue amplifier. The following conditions are assumed in this model:

- The pipelined-SAR ADC is single-ended.
- Capacitors in CDAC are considered without mismatch and the process does not limit the unit capacitor minimum size.

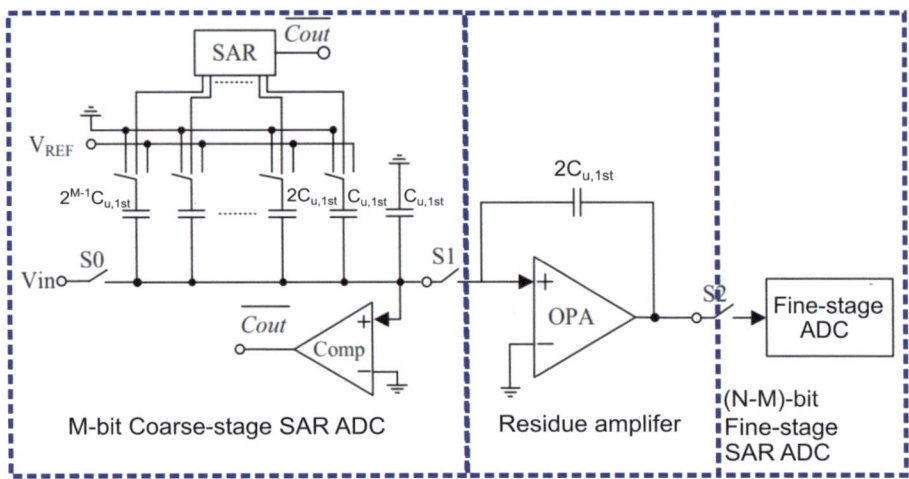

Fig. 6.2 Single-ended pipelined-SAR ADC used in bit-division modeling

- The capacitors in CDAC are fully binary, and the sub-SAR ADCs use a conventional switching scheme.
- The operational amplifier in the residue amplifier is a single-stage amplifier with a common source structure.

- **Sub-ADCs: Unit Capacitor Size in CDAC**
  The size of the unit capacitors in the coarse and fine stages of the CDAC is decisive for the performance of the pipelined-SAR ADC, which has a direct effect on the accuracy. In practice, the minimum value for the unit capacitor is determined by the KT/C noise requirements, the mismatch of the capacitor and the manufacturing process [175]. To simplify the model, only the thermal noise is considered. Large unit capacitor values have lower thermal noise but high switching power and latency. In the CDAC of the coarse ADC, all unit capacitors form the sampling capacitors, which together with the sampling switches generate thermal noise. Since the coarse stage samples the input signal, the thermal noise power needs to meet the overall noise requirements of the ADC. In this case, the root-mean-square voltage of the thermal noise needs to be less than half the LSB of the N-bit resolution. The unit capacitance of the coarse stage can be described as follows:

$$C_{u,1st} = \frac{C_{tot,1st}}{2^M} = 12\,\mathrm{kT}\frac{2^{2N-M}}{V_{FS}^2} \qquad (6.1)$$

6.3 Bit Division Modeling in SAR-Assisted Pipeline ADC

As with the unit capacitors in the fine stage, the sampled noise originates from the residue amplifier. The closed-loop gain of the residue amplifier G_{RA} is 2^M. The feedback factor β_{RA} of the RA can therefore be expressed as follows:

$$\beta_{RA} = \frac{1}{1+G_{RA}} = \frac{1}{1+2^M} \tag{6.2}$$

The output impedance of the residue amplifier R_{out} and the sampled noise power in the fine stage ADC $\overline{v_n^2}$ can then be expressed by:

$$R_{out} = \frac{1}{g_m \beta_{RA}} = \frac{1+2^M}{g_m} \tag{6.3}$$

$$\overline{v_n^2} = 4kT\frac{2}{3}g_m R_{out}^2 \cdot BW \tag{6.4}$$

$\overline{v_n^2}$ relates to the R_{out} of the RA and sampling capacitance or, in other words, to the total CDAC capacitance of the fine stage ADC, $C_{tot,2nd}$. Since the sampled noise only affects the rest of N-M bits, the noise power must be less than the quantization noise power of the fine stage ADC. Therefore, the unit capacitor value of the fine stage $C_{u,2nd}$ can be derived by rewriting $\overline{v_n^2}$:

$$\overline{v_n^2} = \frac{2kT(1+G_{RA})}{3C_{tot,2nd}} = \frac{(\frac{V_{FS}}{2^{N-M}})^2}{12} \tag{6.5}$$

$$C_{u,2nd} = \frac{8kT2^{N-M}}{V_{FS}^2}(1+G_{RA}) \tag{6.6}$$

■ **Sub-ADCs: CDAC Switching Power**

After quantifying the coarse and fine stage unit capacitor value, the switching power of the CDAC can be derived. One of the main sources of energy consumption in SAR ADCs (also in Pipelined-SAR ADCs) is the switching energy of the CDAC, i.e. the energy consumed when charging and discharging the capacitor-based DACs. Assuming that the sub-SAR ADCs use the conventional binary-weighted (CBW) array CDAC with two references (VDD as positive reference and GND as negative reference), the switching energy of the coarse and fine stage of the sub-ADCs can be written according to [176]:

$$P_{sw,1st} \approx f_s 2^M \left\{ \left[\frac{5}{6} - \left(\frac{1}{2}\right)^M - \frac{1}{3}\left(\frac{1}{2}\right)^{2M}\right] V_{ref}^2 \right. \\ \left. - \frac{1}{2}\left(\sum_{i=1}^{M}\frac{D_i}{2^i}\right)^2 - \left(\frac{1}{2}\right)^M V_{ref} \sum_{i=1}^{M}\frac{D_i}{2^i} \right\} \tag{6.7}$$

$$P_{sw,2nd} \approx f_s 2^{N-M} \left\{ \left[\frac{5}{6} - \left(\frac{1}{2}\right)^{N-M} - \frac{1}{3}\left(\frac{1}{2}\right)^{2(N-M)} \right] V_{ref}^2 \right.$$
$$\left. - \frac{1}{2}\left(\sum_{i=1}^{N-M} \frac{D_i}{2^i}\right)^2 - \left(\frac{1}{2}\right)^M V_{ref} \sum_{i=1}^{N-M} \frac{D_i}{2^i} \right\} \quad (6.8)$$

■ **Sub-ADCs: Comparator Power**

In addition to the CDAC power, the comparator is another important source that contributes to power consumption. The comparator is a block between the analog domain and the digital domain. The dynamic latch shown in Fig. 6.3 is often used as a comparator due to its high energy efficiency and speed [177]. It only consumes dynamic current when the current path is formed between the supply and ground, i.e. in the reset and regeneration phase [178]. The differential outputs are reset to the supply voltage before the comparison. During the regeneration phase, the outputs are discharged with a differential current that relates to the inputs. Both the input pair and the cross-coupled inverter pair provide the gain. To simplify the estimation, a fixed current is assumed during the regeneration phase to provide the transconductance. The transconductance can be derived from the noise requirements. Thanks to the pipeline structure, the coarse and fine ADCs only need to fulfill the resolution of their own sub-ADCs. In the coarse stage, it is assumed that the noise power of the comparator is equal to the quantization noise of the M-bit ADC. Then the capacitor load of the comparator can be described as follows:

$$P_{n,cmp1st} = \frac{2}{3}\frac{kT}{C_{n,cmp1st}} = \frac{1}{4}\frac{(\frac{V_{ref}}{2^M})^2}{12} \quad (6.9)$$

$$C_{n,cmp1st} = \frac{32kT2^{2M}}{V_{ref}^2} \quad (6.10)$$

Next, the output time constant of the comparator $\tau_{out,cmp1st}$ can be expressed in:

$$\tau_{out,cmp1st} = \frac{C_{n,cmp1st}}{g_{m,cmp1st}} \quad (6.11)$$

where $g_{m,cmp1st}$ is the total transconductance of the comparator. With an M-bit SAR ADC, the comparator should be able to detect the LSB difference within the regeneration time, i.e. the comparator should be able to amplify an input voltage with LSB value to a full reference voltage within the regeneration time. In this model, the duty cycle of the master clock f_{clk} is 50%, which means that both the reset time and the regeneration time are $1/2 f_{clk}$.

$$V_{out,cmp1st} = V_{in,cmp1st} e^{\frac{t_{reg}}{\tau_{out,cmp1st}}} \quad (6.12)$$

Consider the minimum input of the M-bit comparator, (6.12) can be rewritten as:

6.3 Bit Division Modeling in SAR-Assisted Pipeline ADC

$$V_{ref} = 2^{-M} V_{ref} e^{\frac{g_{m,cmp1st}}{2f_{clk}C_{n,cmp1st}}} \tag{6.13}$$

Then, the $g_{m,cmp1st}$ can be derived. To further derive the regeneration power of the comparator, we assume that the transconductance is provided by a transistor in the saturation region, which means that $gm \approx \frac{2I_D}{V_{dsat}}$. V_{dsat} is the overdrive voltage and is equal to $V_{gs} - V_{th}$.

$$g_{m,cmp1st} = 2(\log 2) M f_{clk} C_{n,cmp1st} \tag{6.14}$$

$$P_{cmp1st,reg} = (\log 2) M f_{clk} C_{n,cmp1st} V_{ref} V_{dsat} \tag{6.15}$$

In the reset phase, the reset power comes mainly from the capacitive output load:

$$P_{cmp1st,rst} = 2 f_{clk} C_{n,cmp1st} V_{ref}^2 \tag{6.16}$$

The total comparator power of the coarse and fine stage sub-ADCs can be described as follows:

$$P_{cmp1st} = C_{n,cmp1st} V_{ref} f_{clk} (ln2 \cdot M V_{dsat} + 2V_{ref}) \tag{6.17}$$

$$P_{cmp2nd} = C_{n,cmp2nd} V_{ref} f_{clk} (ln2 \cdot (N-M) V_{dsat} + 2V_{ref}) \tag{6.18}$$

■ **Sub-ADCs: SAR Logic**
The SAR logic can be built with $2N$ D-type flip-flops (DFF), as shown in Fig. 6.4a for an N-bit SAR ADC [179, 180]. A transmission gate DFF, as shown in Fig. 6.4b, consists of two cross-coupled inverter pairs and four transmission gates [181]. In this case, we can assume that the load of a DFF consists of eight inverters. In this model, leakage is ignored and only the dynamic power is considered for simplicity. The power consumption of the coarse and fine stages of the SAR logic can be written as follows:

$$P_{logic,1st} = 16M^2 \alpha f_s C_{inv} V_{ref}^2 \tag{6.19}$$

$$P_{logic,2nd} = 16(N-M)^2 \alpha f_s C_{inv} V_{ref}^2 \tag{6.20}$$

where C_{inv} is the parasitic input capacitance of an inverter. α is the total switching portion of the SAR logic and is assumed to be 0.5.

■ **Residue Amplifier**
The residue amplifier is used to amplify the residue voltage from the coarse-stage CDAC and pass the output voltage to the fine-stage CDAC, which requires the settling error to be less than half the LSB of the fine-stage.

$$V_{ref} e^{-\frac{T_{set}}{\tau_{amp}}} = 2^{-(N-M)} V_{ref} \tag{6.21}$$

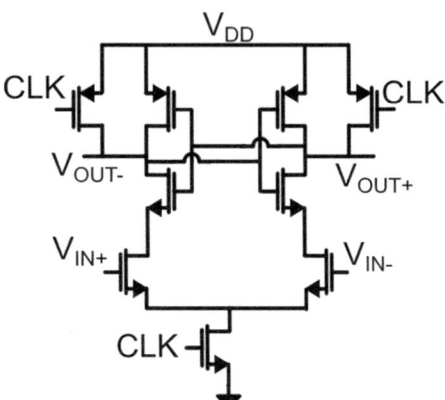

Fig. 6.3 Dynamic latch comparator (*Source* Patel and Veena [177])

where the output time constant of the residual amplifier τ_{amp} relates to the transconductance of the amplifier $g_{m,amp}$ and the total capacitance of the fine stage CDAC from Eq. (6.6). The current consumption of a single amplifier with common source structure and the required settling time can be described as follows:

$$I_{set} = \frac{V_{dsat} g_{m,amp}}{2} = \frac{(N-M)(1+G)C_{tot,2nd} V_{dsat} \ln 2}{2 T_{set}} \quad (6.22)$$

As for the analysis of large signals, the minimum slew rate, which refers to the rise time T_{slw}, can be expressed as follows:

$$I_{slw} = \frac{V_{ref} C_{tot,2nd}}{T_{slw}} \quad (6.23)$$

In this model, we assume that the $T_{slw} = T_{set} = \frac{1}{2 f_s}$. Then the total power consumption of the residue amplifier is

$$P_{amp} = f_s V_{ref}^2 C_{tot,2nd} \left[\frac{V_{dsat}}{V_{ref}} (N-M)(1+G) \ln 2 + 2 \right] \quad (6.24)$$

From the above derivation, we can now deduce the total power consumption of a two-stage pipelined-SAR ADC P_{all}.

$$P_{all} = P_{sw,1st} + P_{cmp,1st} + P_{logic,1st} + P_{sw,2nd} + P_{cmp,2nd} + P_{logic,2nd} + P_{amp} \quad (6.25)$$

A Matlab model was developed to show the relationship between the power consumption of the ADC and the bit division. The simulation results are shown in Fig. 6.5. In this model, the overall accuracy is 14 bits and the number of bits in the first stage changes from 2 to 12 bits. Figure 6.5 (left) shows the absolute power of the three ADC parts. The power of the

6.4 CDAC Switching Power Analysis

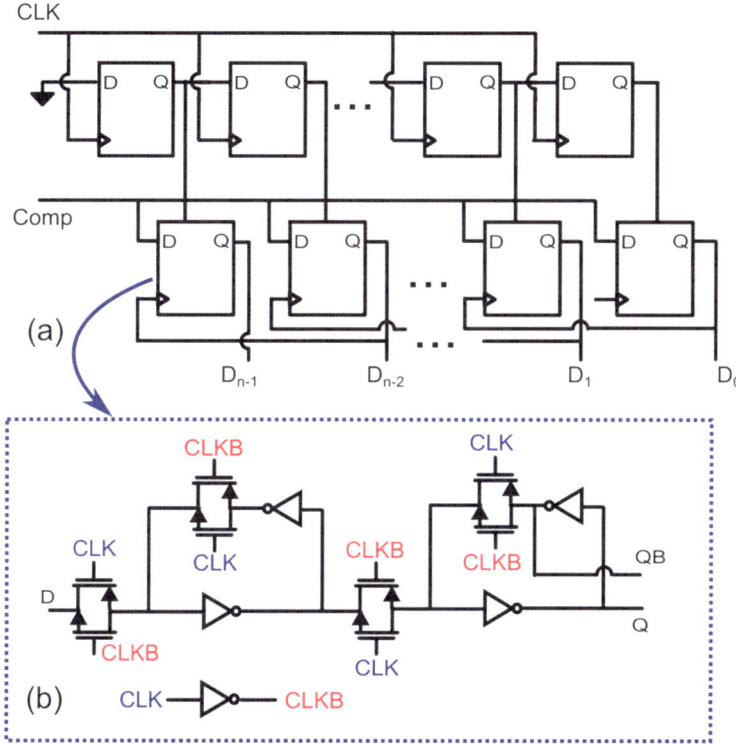

Fig. 6.4 Conventional SAR logic built by D-type flipflops: **a** $2 \times N$ flipflops in SAR logic, **b** transmission-gate DFF

coarse stage increases while the bits in the first stage increase. However, the power of the fine stage and the residue amplifier decreases as more bits are processed in the first stage. Less amplification accuracy is required for the residual amplifier when fewer bits are processed in the fine stage. In other words, an amplifier with a lower gain bandwidth is required. As a result, a U-shaped curve of the total power is drawn and the minimum value is in the middle. This means that an equal number of bits in the coarse and fine stages results in optimum power efficiency.

Figure 6.5 (right) shows the relative share of the individual contributions, i.e. the switching power, the comparator power and the logic power of the individual stages as well as the residue amplifier power. In the two extreme cases, where most of the bits are converted in the first or second stage, the comparator consumes the most power. The reason for this is that the offset and noise of the comparator must be sufficiently small to achieve its accuracy. If an equal number of bits are converted in the first and second stage, the switching of the capacitor bank and the residual amplifier dominate the power consumption.

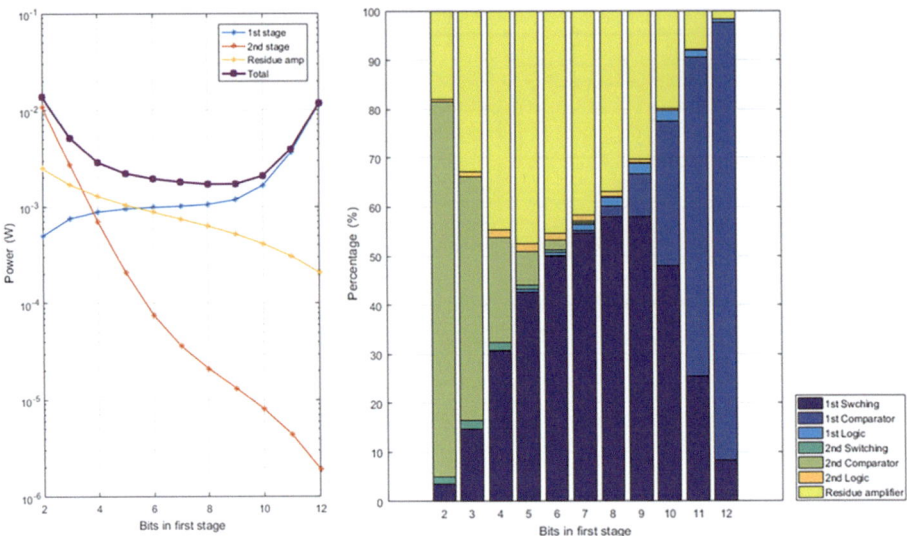

Fig. 6.5 Bit division analysis for a pipelined SAR ADC in Matlab: (left), absolute power of each sub-blocks (right) the percentage share of the individual contributions

In summary, it can be said that the same bit split between the coarse and fine stages results in the optimum total power consumption. Further power reduction can be achieved by improving the power efficiency of the residue amplifier and the switching scheme.

6.4 CDAC Switching Power Analysis

The previous section has shown that equal bit splitting between the coarse and fine stages in a pipelined SAR ADC can achieve optimal power efficiency. However, the CDAC switching energy is one of the main sources of total power consumption from Fig. 6.5 (right). In this section, the switching energy is further investigated with different switching schemes.

Different switching schemes are analyzed to minimize the switching power with minimum design complexity. All switching schemes are discussed with a fully differential SAR ADC to mimic the real switching case. Figure 6.6 demonstrates all possible code paths in a 3-bit SAR ADC with the conventional switching scheme [182]. It can be seen that the switching scheme requires two reference voltages and the first comparison starts after switching phase 1, which consumes the most energy. In addition, Fig. 6.6 shows that the upper switching path consumes much more energy than the lower path because the capacitor reference of each bit can be switched twice depending on the comparator result, which means that the switching energy is highly dependent on the input value.

6.4 CDAC Switching Power Analysis

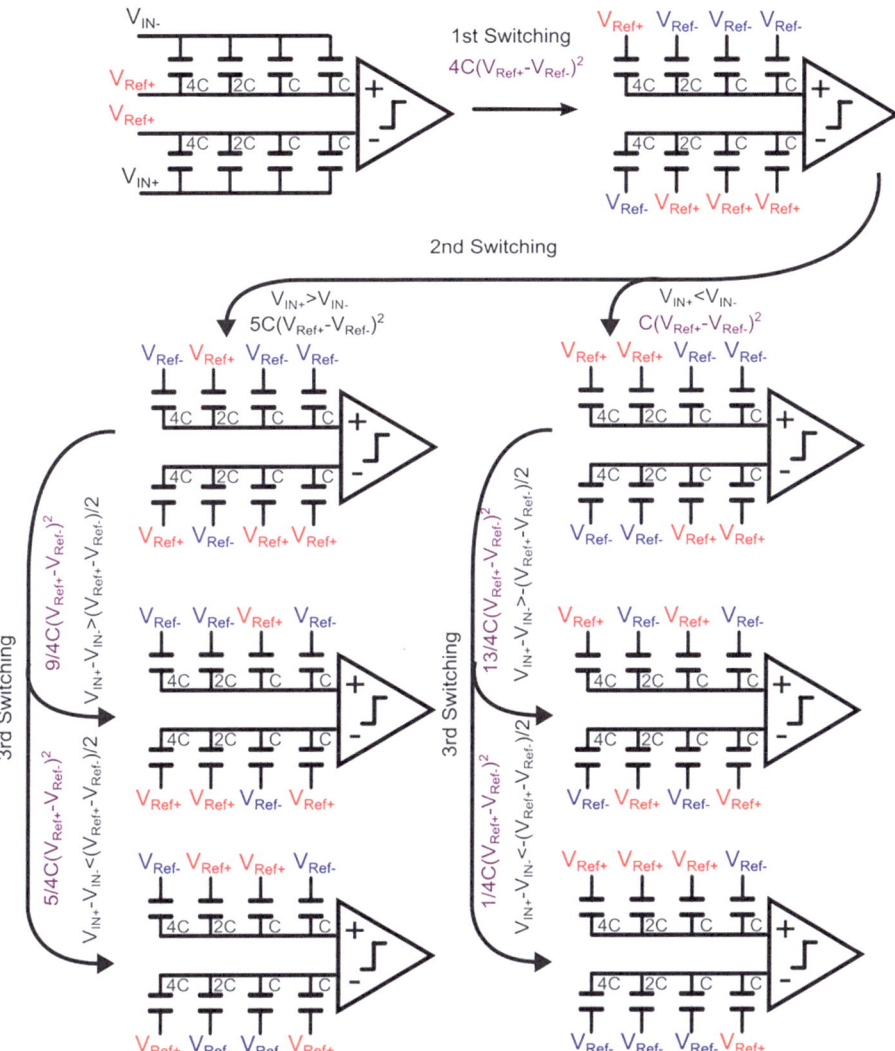

Fig. 6.6 Conventional switching sequence in SAR ADC

The monotonic switching scheme proposed in [183] processes the first comparison directly after sampling without any switching, which means that no energy is consumed before the first comparison. The direct comparison after sampling has another advantage: the MSB capacitor is no longer needed, which saves the CDAC area. Each bit reference only switches once after the comparison, which further reduces the switching energy. Based on the monotonic switching scheme, merged capacitor switching in [184] introduces another voltage reference (common mode reference $V_{cm} = 0.5 V_{refp}$) into the CDAC to further reduce

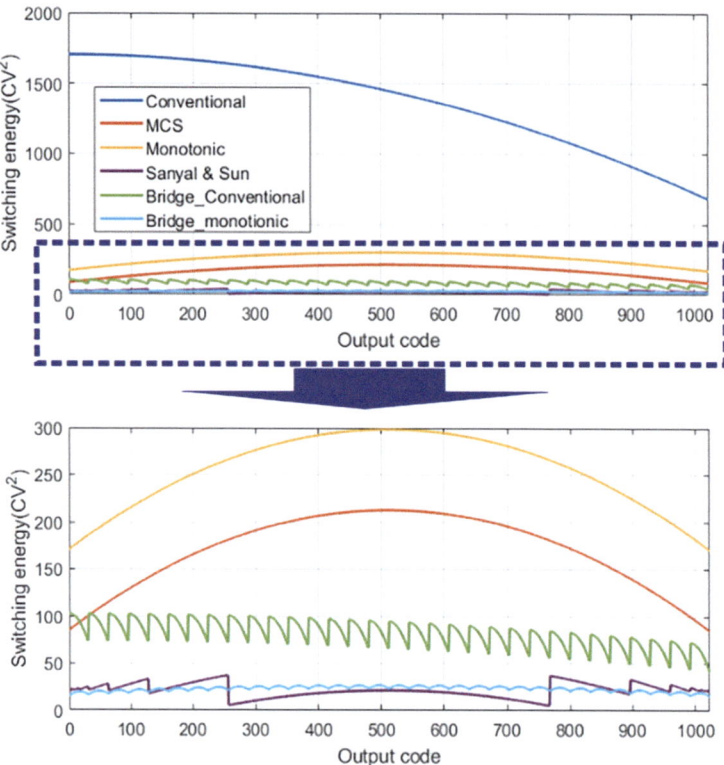

Fig. 6.7 Switching energy with different switching schemes versus output codes for a 10-bit SAR ADC

the switching energy. Due to V_{cm}, the change in the reference voltage on the capacitor is reduced from V_{Vref} to $0.5V_{Vref}$ with each switching, which saves 50% energy compared to monotonic switching. In addition to the previous methods, several works use multiple reference voltages on the MSB capacitors to further reduce the switching energy [185, 186]. However, multiple reference voltage connections with switches on MSB capacitors require additional logic processes and reduce the matching between the capacitors in the CDAC. SAR ADC with a bridge capacitor is another effective way to reduce the total capacitance in the CDAC and the switching energy [187, 188]. However, it requires good matching of the bridge capacitor to achieve sufficient linearity.

A detailed Matlab model was created to analyse all the above switching methods with a 10-bit SAR ADC resolution, and the switching energy versus the output codes is shown in Fig. 6.7 and Table 6.1. It can be seen that conventional switching schemes consume the most energy compared to other methods. The switching energy decreases with the output codes, as switching is performed twice for each capacitor reference. Monotonic switching reduces the energy consumption during switching by more than 80% compared to conventional

6.5 CDAC Capacitor Mismatch and Calibration

Table 6.1 Switching energy comparison table

Switching scheme	CDAC bank	Avg SW energy (CV^2)	Energy saving (%)	#Cu	Vcm required	Comment
Coventional	Binary	1363.3	–	2^{N+1}	No	–
Monotonic	Binary	255.5	81.2	2^N	No	–
MCS	Binary	169.4	87.6	2^N	Yes	–
Sanyal & Sun	Binary	21.3	98.4	2^{N-1}	Yes	Special SAR logic
Conventional	Bridge	82	93.9	$4 \cdot 2^{N/2}$	No	Mismatch limitation
Monotonic	Bridge	23	98.3	$3 \cdot 2^{N/2}$	No	Mismatch limitation

switching schemes. By introducing a common mode reference, Merged Capacitor Switching (MCS) further reduces switching energy. For fully binary SAR ADCs, the switching scheme of [185] saves more than 97% of the switching energy compared to the conventional switching scheme. SAR ADCs with bridge capacitors, which are non-binary ADCs, always have excellent switching energy even with the conventional switching scheme. Table 6.1 shows that MCS saves more than 85% of the switching energy without introducing other special architecture and logic processes. Therefore, this switching scheme is used in the proposed pipelined-SAR ADC.

6.5 CDAC Capacitor Mismatch and Calibration

When modeling the bit splitting, only the thermal noise is taken into account. However, other constraints, such as the mismatch effects during manufacturing, must also be considered. This section investigates the unit capacitor value in the pipelined-SAR ADC with the mismatch effects and explores the mismatch calibration methods.

6.5.1 CDAC Capacitor Mismatch Limitation

From the previous modeling in the bit division, the minimum capacitor for an M-bit coarse ADC results from the limitation of the thermal noise as follows:

$$C_{u,noise} \geq 12kT \frac{2^{2N-M}}{V_{FS}^2} \qquad (6.26)$$

However, mismatch must also be taken into account, which introduces non-linearity into the ADC and generates harmonics in the output spectrum. The mismatch characteristic is highly dependent on the process and the capacitor type. In a fully binary M-bit CDAC with a unit capacitor C_u, the worst DNL occurs when the middle code flips, and the worst DNL can be expressed with the variation of the unit capacitor σ_{C_u}:

$$V_{DNL,wc} = \sqrt{2^M - 1} \frac{\sigma_{C_u}}{C_u} \frac{V_{ref}}{2^M} \tag{6.27}$$

When the maximum DNL is considered to be less than 0.5 LSB of the total N-bit resolution under 3-sigma conditions, the minimum unit capacitor $C_{u,mis}$ from the mismatch limit can be expressed as:

$$3 \cdot V_{DNL,wc} \leq \frac{V_{ref}}{2^{N+1}} \tag{6.28}$$

$$C_{u,mis} \geq 2^{N-M+1} \sqrt{2^M - 1} \frac{\sigma_{C_u}}{C_u} \tag{6.29}$$

And finally, the minimum capacitance is also limited by the process, which can be expressed in $C_{u,process}$. To summarize, the minimum capacitance of a coarse unit that meets all requirements is:

$$C_{u,1st} \geq max\{C_{u,noise}, C_{u,mis}, C_{u,process}\} \tag{6.30}$$

A similar analysis can also be applied to the capacitors of the fine stage unit. Figure 6.8 shows $C_{u,noise}$, $C_{u,mis}$ and $C_{u,process}$ as a function of bit splitting for a 14-bit pipelined-SAR ADC in a 65 nm whose minimum MOM capacitor value is about 10 fF from the foundry PDK. Of course, the customized capacitor value can be much smaller, but the process and mismatch information is not guaranteed by Foundry. If the coarse stage processes less than three bits, the $C_{u,1st}$ is limited by thermal noise. If the coarse stage processes more than four bits, $C_{u,1st}$ is limited by the process mismatch. For this process, the value is limited by the minimum Metal-oxide-metal capacitor of the PDK if the coarse CDAC is more than 10 bits. For a 14-bit pipelined-SAR ADC, the most power-efficient bit division is 7-7 between the coarse and fine stages. In this case, the minimum capacitance C_u increases from 70 to 260 fF when the mismatch effects are taken into account, which also leads to a tripling of the total CDAC. Nevertheless, the total capacitance in the CDAC is preferred to be as small as possible in order to achieve a fast sampling period, a lower load on the front buffer and a smaller chip area. Therefore, calibration of the CDAC mismatch is required to minimize the unit capacitance.

6.5 CDAC Capacitor Mismatch and Calibration

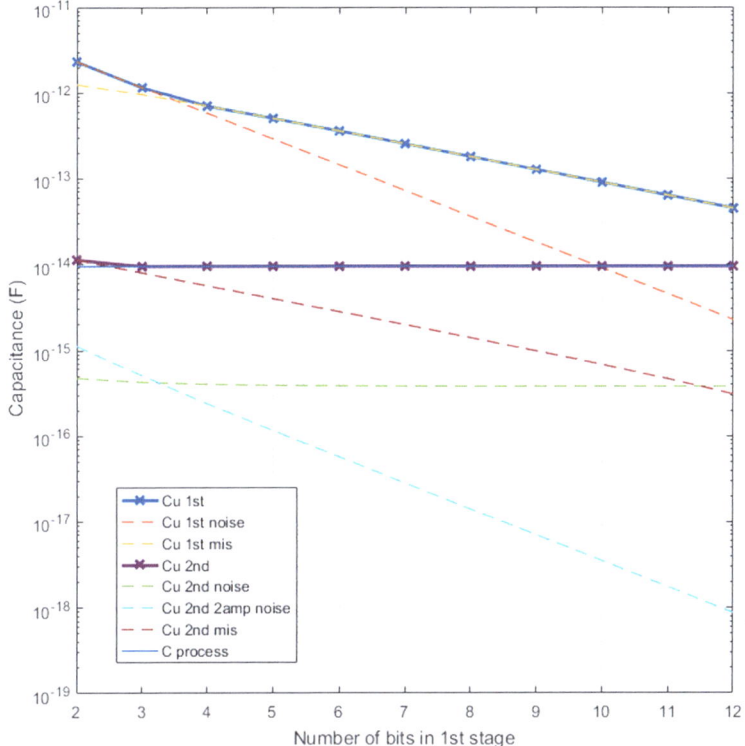

Fig. 6.8 Capacitor size limitations due to thermal noise, mismatch and process

6.5.2 Previous CDAC Calibration Methods

The previous section demonstrated that capacitor mismatch in the coarse-stage CDAC is the primary limitation for pipelined-SAR ADCs, significantly degrading linearity when the resolution exceeds 12 bits [189, 190]. Increasing the size of the unit capacitor can alleviate the mismatch effect in the coarse-stage CDAC. However, this approach leads to a larger sampling capacitor, resulting in a lower sampling rate and higher switching power consumption.

To address the degradation caused by CDAC mismatch, several calibration methods have been proposed, which can be broadly classified into foreground and background calibration techniques as shown in Fig. 6.9. Background calibration, shown in Fig. 6.9a, operates without interrupting the analog-to-digital conversion process, enabling the tracking of long-term variations [191–194]. However, this method imposes additional input requirements and demands significant computational resources. For example, the histogram-based calibration approach in [195] calibrates the mismatch in the background but requires a large dataset to achieve accurate results. In [193], a dithering sequence signal is added to the input to calculate the bit-weight transfer function. Meanwhile, [194] calibrates capacitor mismatch

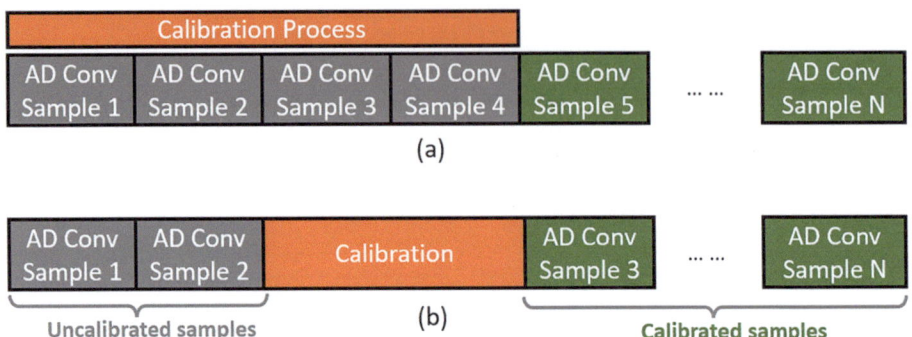

Fig. 6.9 CDAC calibration classification: **a** background calibration, **b** foreground calibration

by analyzing specific input voltage patterns. Nevertheless, if the input signal rarely reaches the required voltages for calibration, the process can be excessively time-consuming or fail to complete. Foreground calibration, shown in Fig. 6.9b, on the other hand, interrupts the analog-to-digital conversion process and is typically performed during idle periods, at power-up, or after significant environmental changes [196, 197]. Since it temporarily halts ADC operation, minimizing the calibration duration is critical to avoid prolonged interruptions.

To overcome these challenges, an advanced calibration method is proposed for pipelined-SAR ADCs, offering reduced digital complexity and faster calibration speed while maintaining high accuracy.

6.5.3 Proposed Calibration Principle

A mismatch calibration process generally involves two primary phases. The first phase is the detection phase, during which the error, such as a capacitive mismatch, is identified. This error is represented either as an electrical quantity, such as a voltage in the analog domain, or as a binary value in the digital domain. The second phase is the correction phase, where the detected error is utilized to adjust and correct the inaccurate output.

6.5.3.1 Capacitor Mismatch Detection

The mismatch detection principle for a pipelined-SAR ADC relies on charge redistribution and CDAC redundancy. A simplified single-ended ADC example illustrates this process in Fig. 6.10, where a coarse-stage 4-bit CDAC includes an additional redundancy bit. If a capacitor mismatch, C_m, exists in BIT(3), the CDAC output voltage differs between equivalent codes like x01000 and x00111, enabling the detection of C_m by measuring this voltage difference. The calibration comprises two steps: Detection Reference Setting (DRS) and Mismatch Sign Extraction (MSE). In DRS, V_{cm} is sent to the residue amplifier and fine-

6.5 CDAC Capacitor Mismatch and Calibration

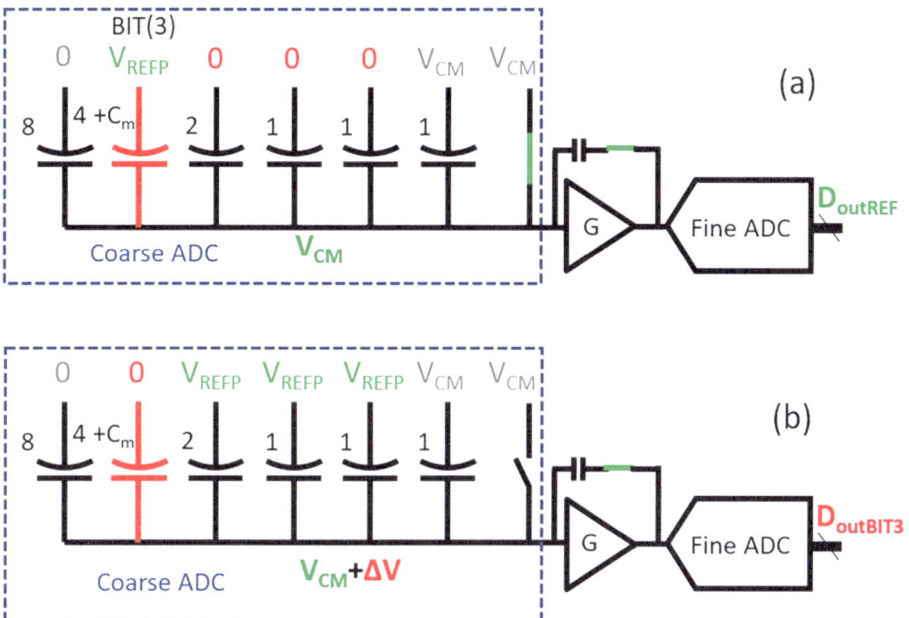

Fig. 6.10 Mismatch sign extraction (MSE) based on the equivalent codes: **a** detection reference setting, **b** mismatch sign extraction

stage ADC, generating a detection reference D_{outREF}. In this way, all kinds of ADC static errors, such as comparator offest and gain mismatches, are includes in D_{outREF}. In MSE, the top plate's voltage is adjusted between equivalent codes (e.g., x01000 to x00111). If C_m exists, the CDAC output shifts to $V_{CM} + \Delta V$, and a new fine-stage ADC output, $D_{outBIT3}$, is generated. The mismatch sign is derived from the difference $D_{outBIT3} - D_{outREF}$, unaffected by static errors in the residue amplifier or fine-stage ADC.

6.5.3.2 Capacitor Mismatch Correction

The correction process involves adjusting the bit capacitor by adding or subtracting a correction capacitor based on the sign of C_m. Figure 6.11 illustrates an example of a 4-bit fully differential CDAC with one redundancy capacitor (only the switches on the positive side are depicted). Each bit capacitor is paired with a Correction Capacitor Bank (CCB), and the figure specifically shows the CCB for BIT(3). The total capacitance of each CCB is determined by the 3-sigma variation of the capacitors and is dependent on the technology used. The unit calibration step, C_{clb}, significantly impacts the calibration performance. Smaller C_{clb} values improve linearity but require longer calibration times and greater digital complexity. If no calibration is performed or the absolute value of C_m is smaller than C_{clb}, the entire CCB follows the switching of the dummy capacitor in the CDAC, as shown

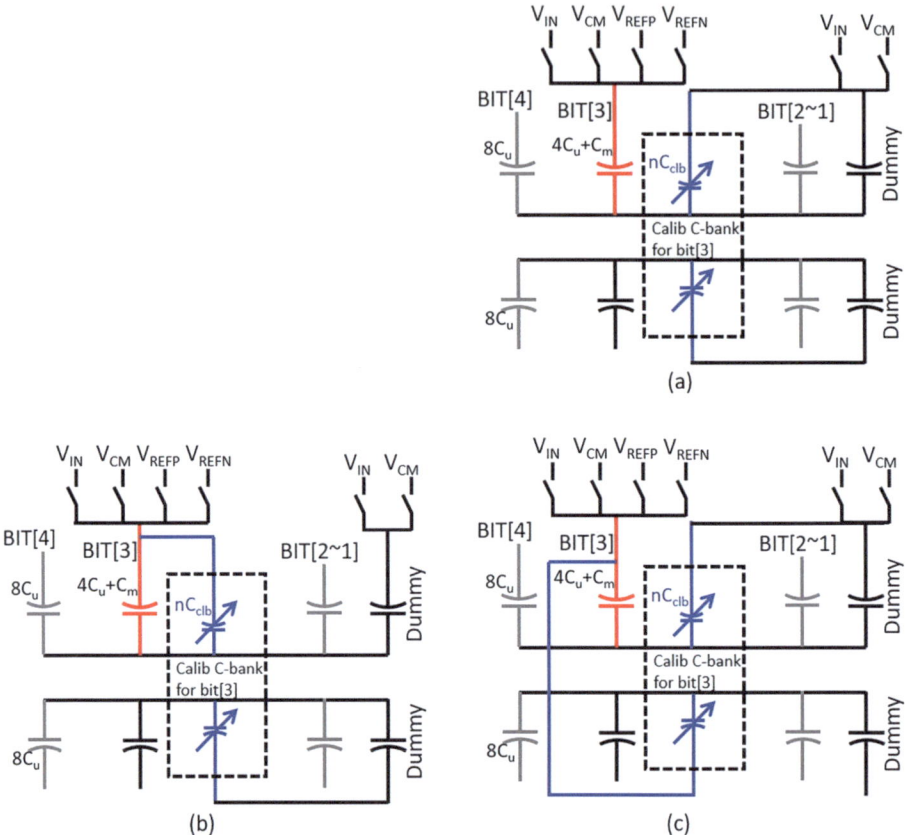

Fig. 6.11 Capacitor mismatch correction **a** no calibration is performed, **b** add capacitors to bit capacitors, **c** substract capacitors from bit capacitors

in Fig. 6.11a. When C_m is negative, a specific number of C_{clb} on the positive side of the CCB is connected to the positive side of BIT(3), achieving capacitor addition as shown in Fig. 6.11b. Conversely, if C_m is positive, a corresponding number of C_{clb} is connected to the negative side of the CCB to function as a subtracting capacito ras shown in Fig. 6.11c. As C_{clb} is connected either to the dummy capacitors or bit capacitors, the overall capacitance of the CDAC remains unchanged during calibration. Consequently, the gain of the residue amplifier (RA) and the sampling capacitance are not affected by the calibration process.

The detailed design consideration, parameters selection, such as the value selection of C_{clb}, and circuits implementation can be found in the [198]. This section only presents the principle of the proposed calibration method.

6.5.3.3 Proposed Calibration Simulation

To validate the effectiveness of the proposed calibration method, a non-ideal coarse-stage CDAC configuration was simulated within a 100 MS/s 14-bit piplined-SAR ADC. All ADC components, except for the calibration logic, were implemented at the transistor level. The CDAC values were extracted from Monte Carlo simulations, representing the worst-case ENOB degradation within a 3-sigma variation. The DFT spectrum for a 49 MHz sine wave input at 100 MS/s is presented in Fig. 6.12. The results indicate that capacitor mismatch degrades the ENOB and SFDR to 11.7 bits and 76.65 dB, respectively. After applying the proposed calibration technique, significant reductions in the noise floor and odd-order harmonics are observed. The ENOB and SFDR improved by 1.93 bits and 17.02 dB, respectively. Figure 6.13 depicts the INL and DNL before and after calibration. The maximum INL reduced from ± 3.52 to ± 1.02 LSBs, aligning with the spectral analysis results. Accounting for all noise sources in the ADC, the final ENOB and SFDR after calibration were measured at 12.58 bits and 90.18 dB, respectively.

6.6 Radiation Tolerance of the Sub-blocks in SAR-Assisted Pipeline ADC

The previous section discussed the system-level considerations to achieve a power-efficient regular pipelined-SAR ADC. If this architecture is used for the radiation environment, its radiation tolerance and hardening strategy should be further investigated. Based on the concluded trade-offs of the radiation-hard ADC design in the previous chapter, pipelined-SAR ADC structures split the overall resolution of the ADC into two sub-ADCs with low resolution and low speed, which improves the SEE tolerance. However, other analog blocks, such as the RA, also pose a greater challenge to SEE tolerance. To avoid over-hardening in the ADC sub-blocks, a detailed analysis of the sub-blocks is performed in this section to achieve efficient hardening in pipelined-SAR ADCs against TID effects and SEEs.

Table 6.2 provides a summary of the radiation effects and their severity on the sub-blocks of a pipelined-SAR ADC. The effects have been categorized into TID effects and SEEs. For the S&H circuits and CDAC switches, TID iffects induce linearity degradation due to charge loss in the sampled signal. The impact of TID is further influenced by the ADC's sampling speed, as longer conversion times allow for greater charge leakage from the sampled charge [199, 200]. The RA is similarly affected, leading to linearity degradation because the residue amplification relies on switched capacitors. Additionally, TID effects degrade the gain and bandwidth of the amplifier and comparator due to shifts in their operating points, as discussed in the previous section [201, 202]. Another critical area of degradation is the clock generation block; the precision of sampling depends on the clock edge quality. TID effects cause the clock edge to become less steep, introducing additional clock jitter. Finally, TID typically results in increased power consumption and reduced speed in most digital sub-blocks.

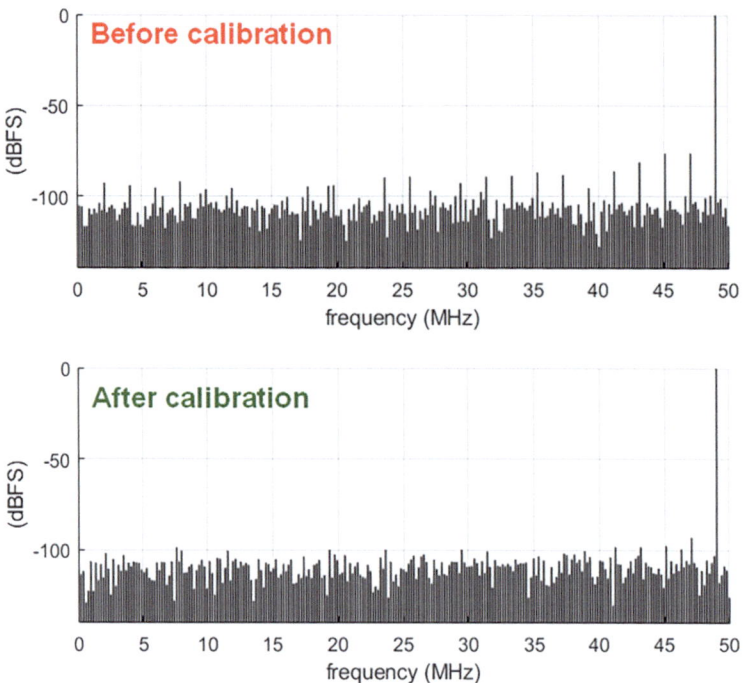

Fig. 6.12 FFT spectra of 14-bit ADC at 100 MS/s for a 49 MHz sine-wave input before and after calibration

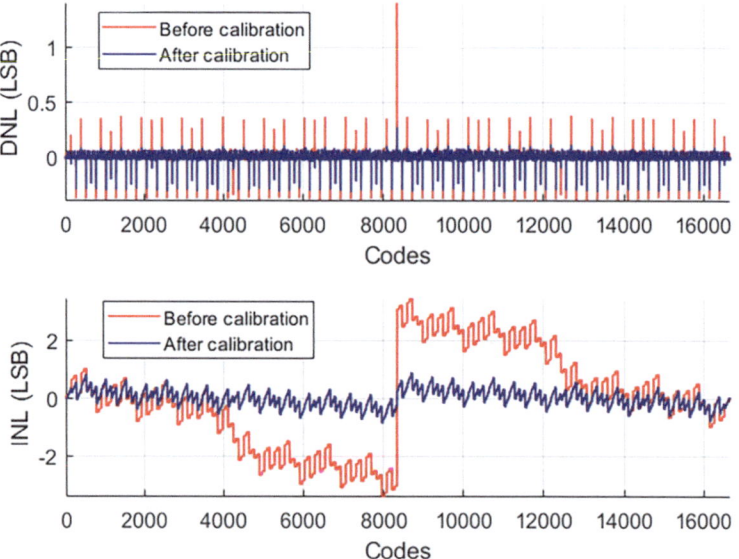

Fig. 6.13 INL and DNL of 14-bit ADC before calibration and after calibration

6.6 Radiation Tolerance of the Sub-blocks in SAR-Assisted Pipeline ADC

Table 6.2 Radiation effects and serenity on sub-blocks of pipelined-SAR ADC

Sub-blocks		TID effects		SEEs	
		Degradation	Severity	Error	Severity
Sub SAR ADC	Sample & hold	Linearity	Low	Single-sample	Low
	CDAC & Switches	Linearity	Low	Single-sample	Low
	Comparator	Gain power	Low	Single-sample	Low
	SAR logic	Speed power	Negligible	Single-sample	Low
Residue amplifier		Gain bandwidth	Low	Multi-sample	High
CLK generation		Jitter	Low	SEFI multi-sample	High
Code alignment		Power	Negligible	Single-sample	Low

In contrast to TID effects, ADCs become increasingly vulnerable to SEEs as technology scales, primarily due to reduced parasitic capacitance and lower supply voltages [70]. From Eq. (5.26), even minor disturbances in analog blocks can propagate into significant output errors. Achieving complete immunity to SEEs would entail substantial development costs and significant reductions in power efficiency. For instance, charge injection into the S&H circuits of high-resolution ADCs could result in deviations of tens to hundreds of LSBs at the output. To ensure complete immunity, the maximum SET indtroduced voltage error needs to be less than half LSB ($\max(\varepsilon_{out}(t)) \leq 0.5 V_{LSB}$), which asks for a large sampling capacitance. In 65 nm technology, a collected charge can reach the pico-Coulomb level when the LET exceeds 60 MeV·cm^2/mg [72]. Under such conditions, achieving full hardening for a 12-bit ADC would require an 8 nF sampling capacitor, which incurs significant chip area, higher power consumption, and a slower sampling frequency. Consequently, complete SEE immunity is not a practical choice for ADCs due to the extensive design complexity and performance trade-offs for analog blocks.

To strike a balance between radiation tolerance, ADC performance, and power efficiency, it is more effective to design the ADC for rapid recovery after an SEE event rather than complete immunity. In the proposed ADC design, the SEE tolerance ensures recovery within a single sampling period, allowing the ADC to produce correct outputs in subsequent periods even if an SEE disrupts a single sample.

Most components in the pipelined-SAR ADC, such as the CDAC, comparator, logic, and code alignment blocks, naturally exhibit single-sample SEE recovery due to being reset by the sub-sampling clock. As a result, SEEs typically affect only one sample, with subsequent outputs recovering once the internal states are reset. However, the clock generation block, which produces the sub-sampling clock, lacks an input reset signal. An SEU in this block can cause an unintended phase shift in the control signals, leading to disordered operation of the RA and sub-ADCs. This results in a SEFI, rendering the ADC output faulty until the clock generation block is manually reset.

Another critical block susceptible to SEEs is the RA. Although it operates as a switched-capacitor amplifier and is periodically reset, its bias circuits lack a reset mechanism. An SEE in the bias circuits can shift multiple bias voltages simultaneously, pushing the RA outside its functional range and compromising the accuracy of the fine ADC stage. Furthermore, these bias circuits are often implemented with low-power current mirrors, making them particularly vulnerable. Errors in the bias circuit can persist for durations ranging from tens of nanoseconds to microseconds, potentially causing multiple output errors depending on the location of the SEE and the bias current. Therefore, it is essential to harden these blocks to prevent extended or repeated errors in the ADC outputs.

6.7 Conclusion

This chapter presents the system-level analysis of a pipelined-SAR ADC and determines the optimal bit division between the coarse and fine stage ADC to achieve optimal power efficiency. Various CDAC switching schemes are also investigated to further minimize the power consumption of the ADC. At an ADC resolution of more than 12 bits, mismatch effects limit the ADC linearity and increase the CDAC sampling capacitance, which degrades the sampling frequency. Therefore, a novel mismatch calibration method is proposed to minimize the harmonics caused by mismatch. At the end of this chapter, all sub-blocks with the TID effects and SEEs are analyzed. TID has a lower impact on the pipelined-SAR ADC in the 65 nm process thanks to process scaling. However, for SEEs, the residue amplifier and some digital circuits, such as the clock generation block, can cause multiple errors or even unrecoverable failure of the ADC. They need to be hardened to increase the overall SEE tolerance.

7 13-Bit High-Performance Radiation-Tolerant ADC

Abstract

This chapter presents a radiation-tolerant 13-bit ADC that was realized in 65 nm CMOS technology and whose development took into account the compromises and considerations from the radiation-tolerant ADC design trade-offs. A Semi-time-interleaved pipelined-SAR structure is used, which provides both power efficiency and SEE tolerance. Each sub-block of the proposed ADC is made radiation tolerant for TID and SEE at both architectural and circuit levels. Through electronic measurements, the prototype ADC achieves 70.79 dB SNDR and 80.26 dB SFDR at the Nyquist input frequency and a sampling rate of 80 MS/s. In addition, TID irradiation tests confirm that the ADC remains unaffected up to 500 krad(Si) and has a robust withstand capability. The ADC has a limited SEE-sensitive range and also recovers quickly from SEE events.

7.1 Introduction

The increasing investment in space exploration and satellite applications has amplified the demand for radiation-hardened electronics with high performance. By November 2021, a large number of small satellites had been deployed into Low Earth Orbit (LEO) as part of the Starlink constellation, with long-term plans aiming for thousands more [203]. Additionally, the Artemis missions led by NASA and ESA have initiated the deployment of scientific instruments and technology experiments on the lunar surface through commercial payload delivery services starting in 2022 [204]. Space missions, including satellite constellations, planetary rovers, and interstellar probes, require advanced electronic systems for data acquisition, processing, and transmission. ADCs and DACs, which are fundamental components in these systems, must maintain high reliability and performance in the extreme radiation conditions encountered in space.

However, achieving radiation tolerance comes with trade-offs, often affecting power consumption, speed, or accuracy, thereby reducing the overall efficiency of ADCs, as analyzed in Chap. 5. At the system level, a widely adopted approach to enhance ADC resilience against radiation is TMR, which employs three identical ADCs along with an output voter. Despite its effectiveness, TMR significantly increases total power consumption and necessitates a highly linear and robust input buffer. Additionally, the simultaneous operation of three ADCs imposes higher demands on reference voltage generation and decoupling capacitors, leading to increased area usage and greater reference noise. An alternative, Double Modular Redundancy (DMR), as discussed in [205], aims to lower power consumption. However, while DMR can detect errors caused by SEEs, it merely discards erroneous data without error correction, still requiring twice the power of a single ADC. Another approach, proposed in [206], utilizes a higher power supply to mitigate radiation-induced degradation. These prior works primarily focus on system-level hardening of ADCs without comprehensively addressing the four key design trade-offs in radiation-hardened ADCs: radiation resilience, power efficiency, speed, and accuracy. However, based on the ADC system-level analysis in Chap. 5, it is possible to improve SEE tolerance with minimal impact on power and speed, for example, by implementing reset mechanisms in both digital and analog circuits. Furthermore, as analyzed in the pipelined-SAR ADC block-level discussion in Chap. 6, only the most vulnerable components, such as amplifier biasing and clock generation circuits, require hardening, as other blocks have a relatively minor influence on overall SEE resilience.

This chapter introduces a 13-bit pipelined-SAR ADC implemented in a 65 nm CMOS process, designed specifically for space applications [207]. The ADC operates at a sampling rate of 80 MS/s, achieving an SNDR of 70.8 dB and an SFDR of 80.3 dB at the Nyquist input frequency. To ensure both high efficiency and robust radiation tolerance, design optimizations were applied across the architectural, block, and transistor levels. With a total power consumption of 13.8 mW, the prototype ADC attains a Walden Figure of Merit of 60.7 fJ/conversion-step, demonstrating efficiency comparable to that of conventional ADCs without radiation-hardening features but with similar performance specifications.

7.2 ADC Top-Level Architecture Selection

As discussed in Sect. 5.6.3, enhancing the radiation tolerance of an ADC can be achieved through resolution and speed decomposition. Therefore, the proposed ADC architecture is developed based on these two principles.

For resolution decomposition, a pipelined structure is a commonly adopted approach, where multiple sub-ADC stages are connected in series via one or more residue amplifiers. In this design, a two-stage pipeline with two sub-ADCs is chosen to balance design complexity and power efficiency. Among the various Nyquist ADC architectures, the SAR ADC is an attractive choice for sub-ADC implementation due to its balanced accuracy and speed, simple structure, low power consumption, and scalability with advanced technology nodes [159,

172, 173]. More importantly, since SAR ADCs primarily consist of a capacitor digital-to-analog converter (CDAC), SAR logic, and a comparator, their predominantly digital nature makes them more resilient to radiation-induced effects. Additionally, capacitors, being passive components, are less affected by ionization effects [114].

For sampling speed decomposition, time-interleaving can be applied to the aforementioned architecture. Consequently, a time-interleaved pipelined-SAR ADC is a promising solution to achieve both high resolution and high-speed sampling while mitigating radiation sensitivity. Prior studies have demonstrated that this architecture delivers outstanding ADC efficiency [162, 174]. However, structural-level enhancements alone are not sufficient for achieving robust radiation tolerance. Additional hardening at the block and circuit levels is essential, as different ADC blocks exhibit varying susceptibilities to radiation effects. Therefore, the next subsection provides a detailed sensitivity analysis of sub-blocks to implement effective radiation-hardening techniques in time-interleaved pipelined-SAR ADCs.

7.3 Proposed ADC Structure with Efficiency Improvement

Conventional two-stage pipelined-SAR ADCs, illustrated in Fig. 7.1a, perform the MSBs and LSBs computations over two sampling periods. The corresponding timing diagram for each sample is shown in Fig. 7.1b. In this process, the coarse ADC (C-ADC) determines the initial MSBs and generates a residue voltage. This residue is then amplified by the residue amplifier (RA) and forwarded to the fine ADC (F-ADC) for LSB computation. The advantages of the pipelined-SAR architecture in terms of power efficiency and radiation tolerance are discussed in Chap. 6.

However, a notable drawback of the conventional pipelined-SAR approach is the incomplete utilization of the RA in each conversion cycle due to its sequential operation, which reduces ADC efficiency. One way to optimize RA utilization is to extend its amplification

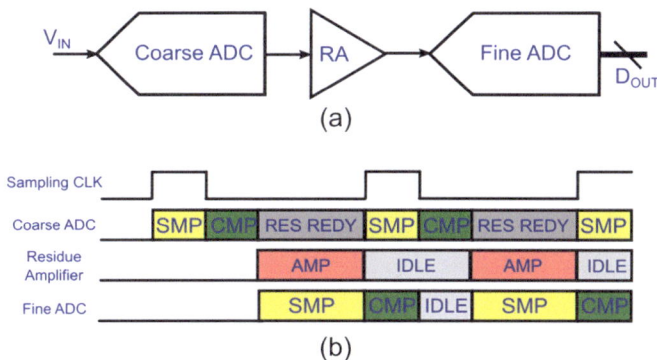

Fig. 7.1 Conventional pipelined-SAR ADC **a** block diagram, **b** timing diagram

phase to maximize its active duration within the sampling period. However, this approach shortens the comparison time available for both the C-ADC and F-ADC. An alternative solution is to power off the RA when not in use to conserve energy. Given the 13-bit resolution and 80 MS/s sampling rate, the RA requires a relatively high open-loop gain (>70 dB) and wide bandwidth (>160 MHz). Implementing rapid power-on/off functionality for the RA incurs additional power consumption and design complexity, making its practical implementation challenging.

A Semi-Time-Interleaved (Semi-TI) pipelined-SAR ADC is introduced to maximize the utilization of the RA and improve power efficiency. As shown in Fig. 7.2, the proposed ADC consists of two pipelined-SAR channels sharing a single RA, ensuring continuous operation by alternating between the two channels. The system operates with a master clock (Φ_{CLK}) of 160 MHz, generating 40 MHz sub-sampling signals ($\Phi_{S1,2}$, $\Phi_{A1,2}$) for each channel and the RA. During each channel's sub-sampling phase (when $\Phi_{S1,2}$ are high), the RA resets and prepares for amplification. Then, the RA serves channel 1 and 2 alternatively. This approach reduces the bandwidth requirement of the C-ADC, F-ADC, and RA, as they only operate at half the sampling frequency ($f_s/2$). The C-ADC, depicted in Fig. 7.3, features a 6-bit resolution with 1-bit redundancy, while the F-ADC achieves 7-bit accuracy with an extra

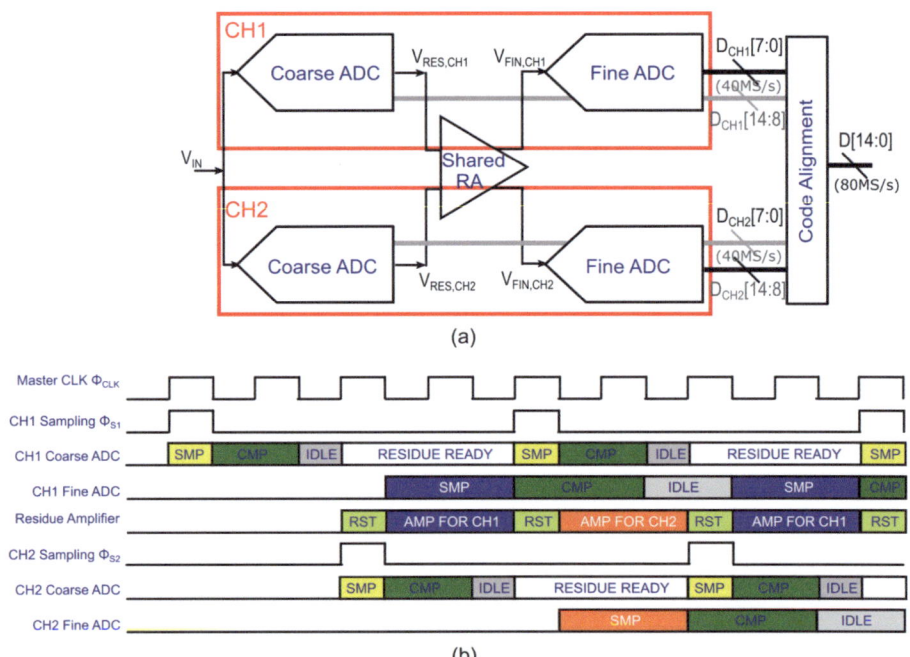

Fig. 7.2 Radiation-tolerant Semi-Time-interleaved pipelined-SAR ADC structure **a** block diagram, **b** and timing diagram

7.4 Design Details of the Key ADC Sub-blocks

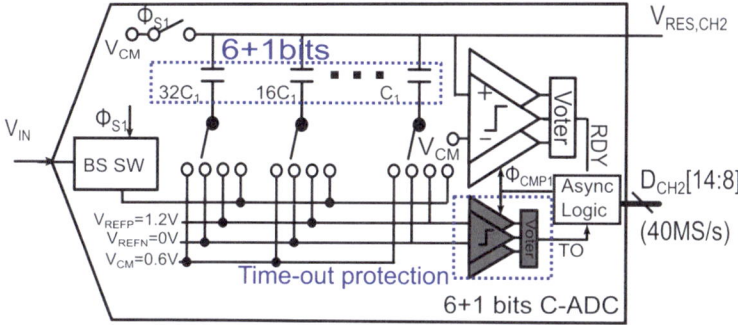

Fig. 7.3 Coarse-stage SAR ADC structure (fine-stage ADC has the similar structure)

interstage redundancy bit. As a result, the RA is designed with a half-scale gain (32.5×), improving error tolerance in the coarse stage.

The Semi-TI architecture offers significant advantages in radiation tolerance. The RA is fully utilized throughout the conversion cycle, eliminating idle periods and enhancing efficiency. Additionally, operating all sub-ADCs at half f_s reduces bandwidth constraints and allows for longer SEE recovery times, as discussed in Eq. (5.25). The lower resolution per sub-ADC further improves SEE resilience, as analyzed in Eq. (5.26). Moreover, asynchronous SAR logic is employed in all sub-ADCs, integrating a time-out protection mechanism to prevent excessive comparison delays that could extend SAR logic cycles. Further details on the time-out protection and comparator design will be discussed in a later section.

7.4 Design Details of the Key ADC Sub-blocks

7.4.1 Shared Residue Amplifier

7.4.1.1 Conventional Residue Amplifier with Auto-Zeroing Offset Cancellation

For the residue amplifier in pipelined-ADC structure, the switched capacitor amplifier is one of the most used structures. It leverages the charge transfer principle using capacitors and switches to achieve amplification without requiring a traditional resistive feedback network, making it highly suitable for integrated circuit implementation. This architecture inherently provides high linearity and precision while being less affected by process variations compared to resistive amplifiers. Additionally, it enables programmable gain by adjusting the capacitor ratios, offering flexibility in various applications. Switched-capacitor amplifiers are also power-efficient and compatible with modern CMOS processes, making them an ideal choice for low-power and high-performance designs, such as pipelined ADCs and residue amplifiers in data converters.

When the operational amplifier is exposed to long term radiation, threshold shift and leakage degradation are introduced by TID effects. However, the degradation of each transistor is not exactly the same. Such difference in a differential pair can results to additional offset of an fully differential amplifier. To make matters worse, this TID-induced offset is changing with total dose and time. As a result, when is op-amp is used in a switch capacitor residue amplifier mentioned above, the varying input offset will finally affect the residue amplification of RA. This offset may saturate the RA output when both amplification gain and the input offset is big. The input offset of the op-amp in the RA can reduce the inter-stage redundancy or even saturate the input of the F-ADC. Auto-Zeroing (AZ) is needed to cancel the input offset [208].

The conventional RA used in pipelined-SAR ADC with Auto-zeroing offset cancellation is shown in Fig. 7.4. The CDACs in the coarse-stage and fine-stage ADC are simplified. The operation typically involves two phases: a sampling phase ϕ_S and an amplification phase ϕ_A. At the sampling phase ϕ_S, the capacitor bank of the coarse stage ADC, which is simplified in a single capacitor $C_{DAC,C}$, is sampling the entire ADC input. The bottom plate of $C_{DAC,C}$ is connected to V_{in}, while the top plate is connected to V_{cm}. At the same time, the RA resets the feedback capacitor C_{fb}. Besides, the op-amp is performing the AZ offset sampling and the offset voltage is sampled on the capacitor C_{az}. In this phase, the fine-stage ADC capacitor bank is disconnected from the RA. When moving to the amplification phase ϕ_A, the coarse-stage ADC provided its residue voltage by connecting the bottom plate of $C_{DAC,C}$ to V_{sw}. V_{sw} is an equivalent CDAC voltage that correlates to the final comparison results of the coarse stage. Then, this residue voltage is amplified by the ratio of $G_{RA} = C_{DAC,C}/C_{FB}$ and the amplified voltage is sampled by the fine-stage ADC.

However, such a structure imposes an instability issue, which later causes the power inefficiency. As explained in the previous paragraph, there are two feedback loops in two

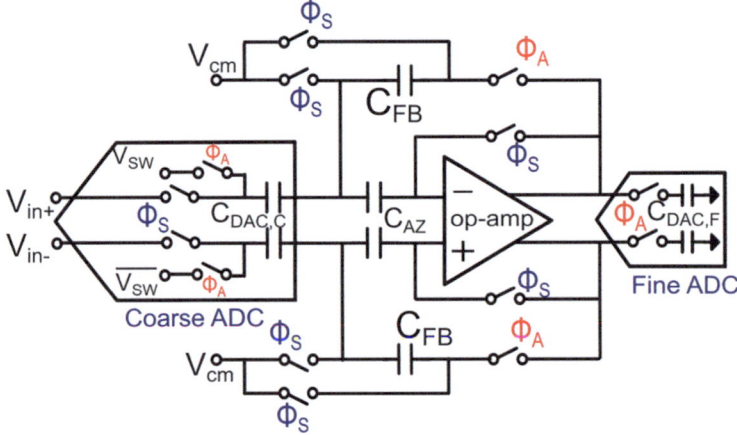

Fig. 7.4 Conventional RA in a pipelined-SAR ADC with Auto-zeroing offset cancellation

7.4 Design Details of the Key ADC Sub-blocks

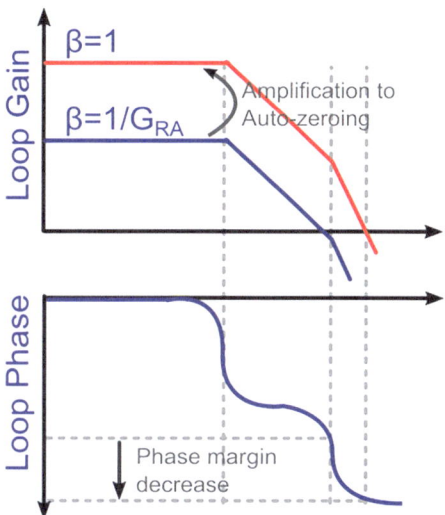

Fig. 7.5 Conventional RA in a pipelined-SAR ADC with Auto-zeroing offset cancellation

operating phases. During the amplification phase ϕ_A, C_{fb} and $C_{DAC,C}$ form the feedback path and result in a feedback factor of $\beta_A = 1/G_{RA}$. On the contrary, in the sampling phase, due to AZ, the $\beta_A = 1$. If we simply assume the loop has two dominant poles, then we can draw a Bode plot of the two loops as shown in Fig. 7.5. One can observe that changing the state of the RA can result in a DC loop gain increase of G_{RA}. The higher the amplification ratio of the RA, or in other words, the higher the coarse-stage ADC resolution, the higher the DC gain change will be. If the C_{az} is not too huge to significantly change the position of the dominant pole and the second dominant pole, the loop stability becomes worse when switching from AZ to amplification mode. In this case, more effort (power, area, BW) has to be paid to stabilize both loops.

7.4.1.2 Proposed Residue Amplifier Structure

The root cause of the power inefficiency of the conventional structure is that each cycle contains both an AZ and an amplification phase. But AZ is not mandatory for each sample since offset change due to the TID effect is slow in space applications. Therefore, the main idea of the proposed structure is to split the two loops, which allows a slow AZ and a fast amplification loop. The proposed RA contains two op-amps operating in a ping-pong scheme, as shown in Fig. 7.6a. For most of the operation, only one op-amp is enabled and serves as amplification for both channels. Thus, all inputs and outputs of the two op-amps are isolated by switches. Before swapping between one op-amp and the other one, this is the period when both op-amps are operating: One op-amp is doing amplification while the other one is performing auto-zeroing calibration. The example case is also annotated by color in Fig. 7.6a.

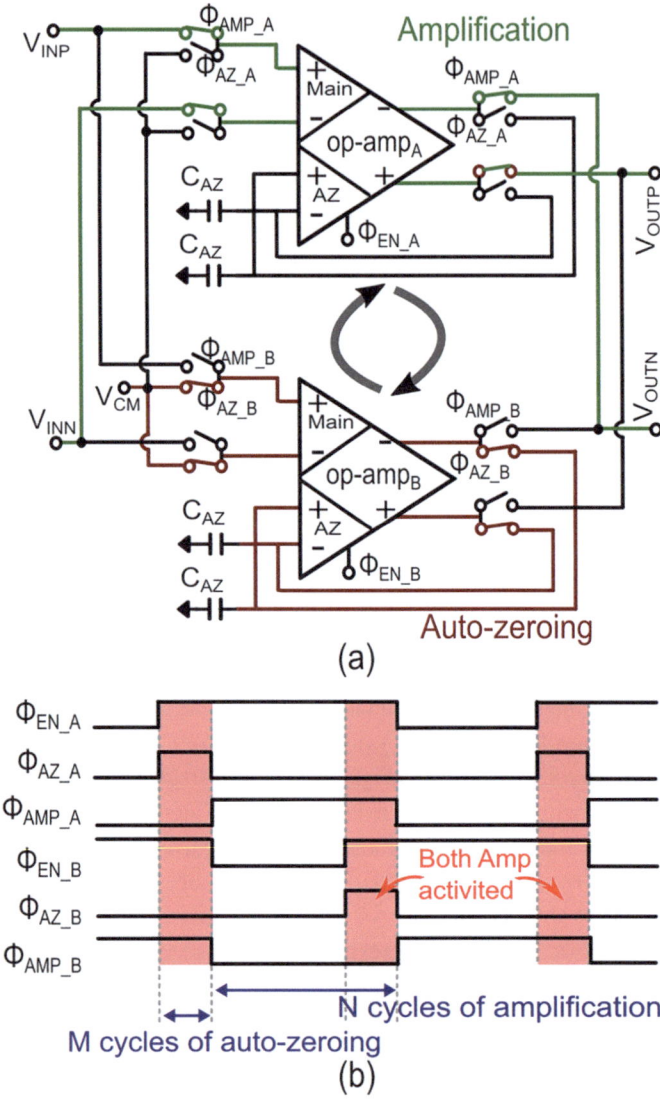

Fig. 7.6 Residue amplifier with two op-amps running in ping-pong scheme **a** block diagram, **b** timing diagram

7.4 Design Details of the Key ADC Sub-blocks

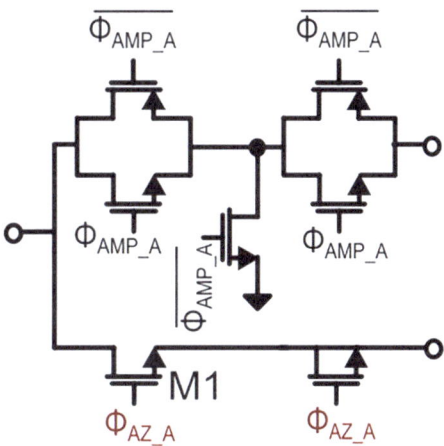

Fig. 7.7 The output switch of RA to avoid coupling between AZ and amplification

The timing operation of the proposed RA is illustrated in Fig. 7.6b. As shown in the timing diagram, AMP_A alternately serves both ADC channels during the initial N samples. At the $(N - M)$th sample, AMP_B is powered on and begins the AZ process, while AMP_A continues amplification. Once AMP_B completes AZ after M samples, it takes over amplification duties from AMP_A, which is then powered down. This approach ensures that both amplifiers are active only during the AZ phase, optimizing power efficiency. The total power consumption of the RA can be approximated as:

$$P_{RA} \approx \frac{2N + 2M}{2N} P_{op-amp} = (1 + \frac{M}{N}) P_{op-amp} \tag{7.1}$$

where P_{RA} represents the total power consumption of the RA, and P_{op-amp} denotes the power consumed by each operational amplifier within the RA. By extending the AZ period across multiple samples, the required bandwidth is reduced, enhancing stability and significantly lowering the power demand of individual op-amps. Additionally, when $N \gg M$, the power overhead of the AZ phase becomes negligible. However, during the AZ phase of one op-amp, the output of the other may inadvertently couple to C_{AZ} via SW_1. To mitigate this interference, instead of a single transmission gate, two transmission gates combined with an NMOS transistor connected to ground are employed, ensuring greater isolation between the two amplifiers, as depicted in Fig. 7.7.

7.4.1.3 SEE Hardening of the Op-Amps

The op-amps in the RA employ fully differential folded-cascode class-AB architectures with gain boosting, as depicted in Fig. 7.8, to achieve high gain and fast settling. The AZ input pair, common-mode feedback, and biasing circuits are not shown for simplicity. Since AZ is not performed in every cycle, as discussed earlier, the op-amp is reset by shorting its positive

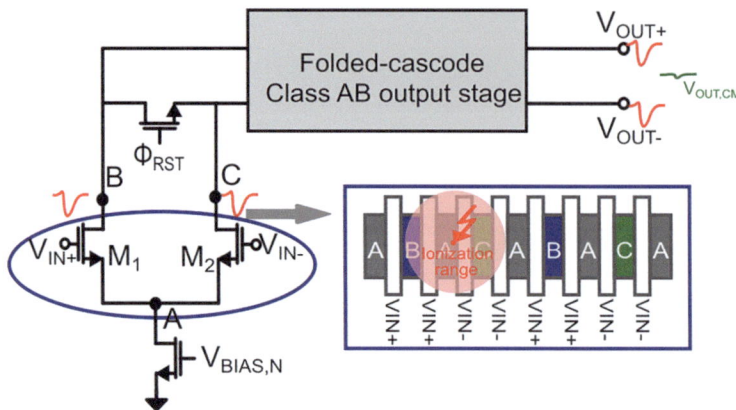

Fig. 7.8 Fully differential folded-cascoded class-AB amplifiers inside the RA (CMFB amplifier and AZ input pair are not shown)

and negative branches. The designed amplifier achieves an AC open-loop gain exceeding 90 dB and a closed-loop bandwidth greater than 280 MHz.

A notable design feature is the interleaved centroid layout used for all differential pairs within the op-amp, as indicated in Fig. 7.8, though dummy transistors are not shown. This layout technique effectively reduces mismatch and process variation effects while enhancing resilience against SEEs [209]. When an energetic particle strikes the active region of a transistor, such as the input differential pair in Fig. 7.8, a funnel-shaped region forms, generating electron-hole pairs. The charge collection radius in the semiconductor active layer depends on material properties and ion characteristics, potentially reaching micrometer levels [141]. Due to the interleaved layout, the collected charge is distributed between both input transistors, reducing the differential output impact. Since charge sharing from an SET primarily results in common-mode variations, the differential operation of the amplifier remains largely unaffected.

Compared to the main amplification branches of the operational amplifier, the biasing circuits exhibit greater susceptibility to SEEs. An example of the biasing circuits is shown in Fig. 7.9. If a SET occurs at transistor M1, the impact of a heavy ion may induce an unintended current flow from the drain to the bulk of M1. Consequently, this disturbance propagates through the biasing network, affecting all bias voltages $V_{BIAS1\sim3}$. This instability can persist for multiple samples, leading to erroneous ADC outputs.

To mitigate these effects, several hardening techniques are implemented to the bias circuits as shown in Fig. 7.10. First, the biasing circuits are segmented into multiple unit blocks, each generating a bias voltage that is subsequently recombined through series resistors. Therefore, a low pass filter is formed by the resistor and node parasitic capacitor in each slice. Additionally, the biasing blocks are strategically placed with sufficient spacing in the layout to minimize charge sharing from a single SET. These design strategies significantly

7.4 Design Details of the Key ADC Sub-blocks

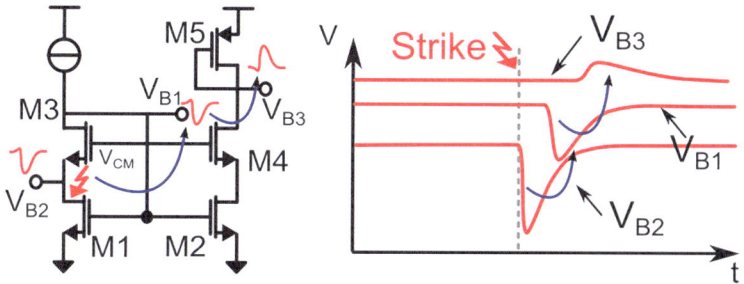

Fig. 7.9 Biasing circuits before and after hardening for SEEs

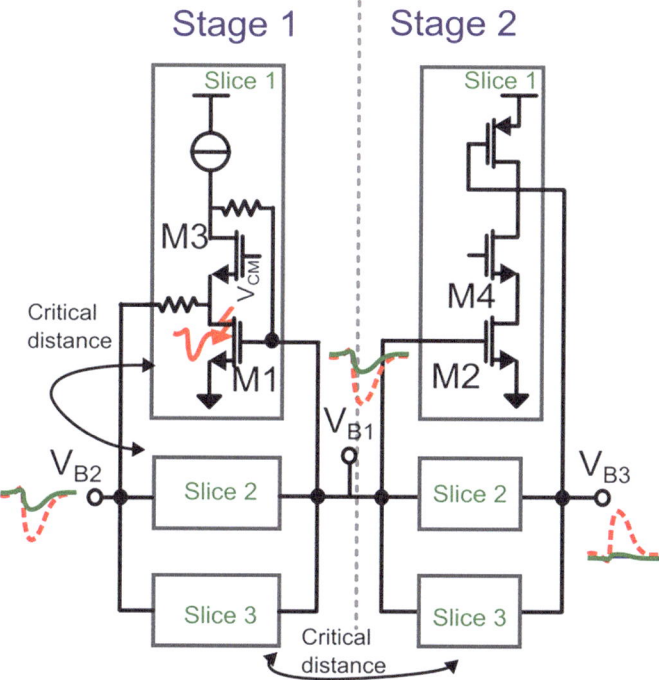

Fig. 7.10 SET simulation for all transistors in the bias circuit

enhance the robustness of the biasing circuit against transient disturbances. A transient simulation is shown in Fig. 7.11 to compare the bias voltage as well as the op-amp output before and after hardening. It can be seen that the correlation between different bias voltages is largely reduced and results in lower error at the op-amp outputs in both amplitude and duration.

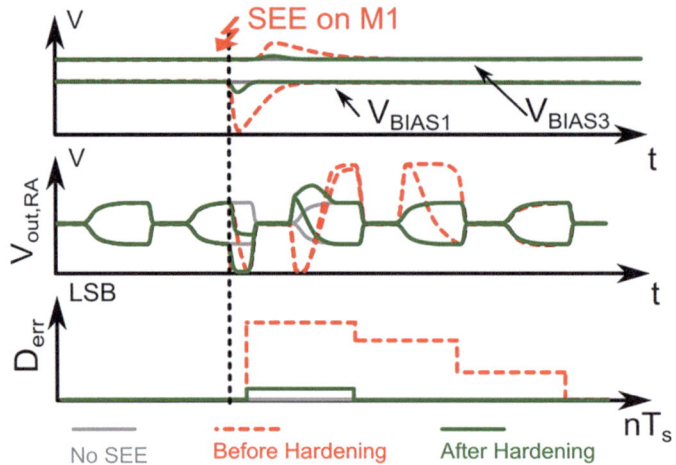

Fig. 7.11 SET simulation for all transistors in the bias circuit

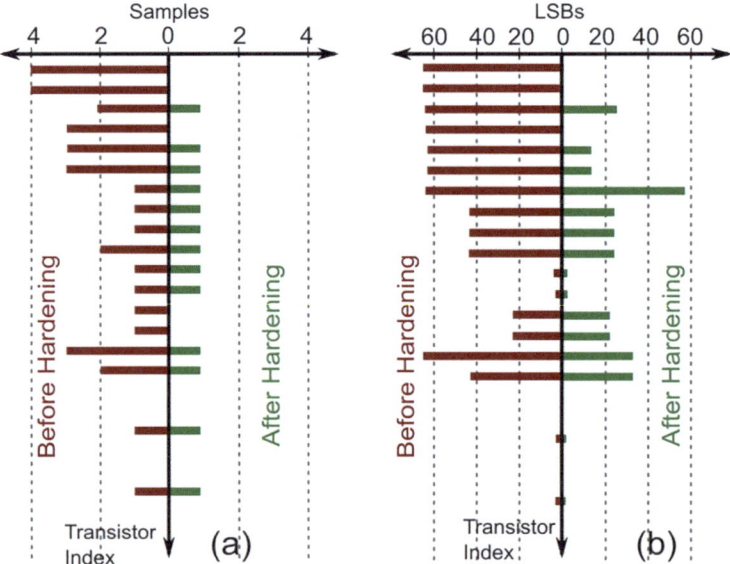

Fig. 7.12 SET simulation for all transistors in the bias circuit

All transistors within the biasing circuits with the proposed hardening structure are evaluated using an SET current model [72] at a LET of 60 MeV · cm^2/mg. The observed error in the LSB is significantly reduced, and, crucially, the ADC outputs recover to within ± 0.5 LSB within a single sample as shown in Fig. 7.12.

7.4.2 Clock Generation

The ADC operates with a 160 MHz fully differential sinusoidal input signal and generates the sub-sampling clock signals $\Phi_{S1,2}$ along with the RA control signals $\Phi_{A1,2}$ and Φ_{RST}. The clock generation block, illustrated in Fig. 7.13a, processes the input sinusoidal signal through an input buffer. This buffer, composed of inverters and cross-coupled latches, amplifies the sinusoidal waveform into a square-wave clock signal Φ_{CLK} with a sharp falling edge and a 50% duty cycle. Subsequently, a frequency divider and logic gates generate two raw 40 MHz sub-sampling clocks ($\Phi_{S1,R}$ and $\Phi_{S2,R}$), which maintain a 180° phase shift.

To minimize clock skew between $\Phi_{S1,R}$ and $\Phi_{S2,R}$, an adjustable MOS capacitor-based calibration mechanism is implemented as illustrated in Fig. 7.14. Each calibration block consists of 64 units, controlled via a 6-bit digital code. Within each unit, the gates of NMOS and PMOS transistors are interconnected and tied to the target clock signal, while their drain and source terminals are shorted and controlled by the calibration code. By varying the gate-source voltage, the transistor channel capacitance can be modulated, achieving a fourfold variation in capacitance. This configuration enables a calibration resolution of 17.5 fs per step, with an overall calibration range of ± 1.12 ps between channels. Following calibration, the final sub-sampling clocks Φ_{S1} and Φ_{S2} exhibit a clock jitter below 40 fs and a clock

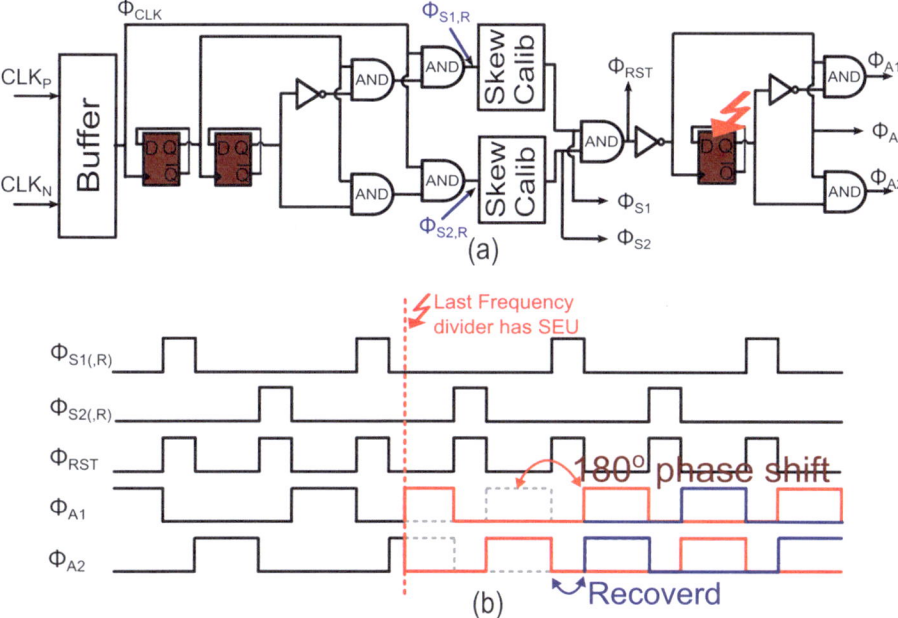

Fig. 7.13 Clock generation block in proposed ADC **a** block diagram, **b** timing diagram

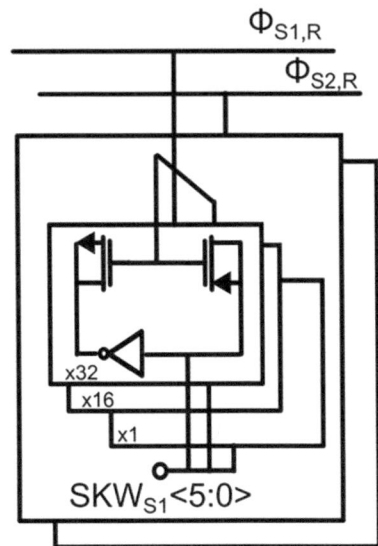

Fig. 7.14 Skew calibration for two sampling clock

skew of less than 17.5 fs. These calibrated clock signals are then used for sampling and controlling the RA.

It is worth noting that the clock generation module includes three frequency dividers. Conventionally, a frequency divider may use a single D flip-flop without a dedicated reset input [210]. However, such an arrangement is susceptible to functional errors if an SEU alters the state of one of the DFFs. As shown in Fig. 7.13b, an SEU affecting the final stage of frequency division can cause an unexpected 180° phase shift in the control signals Φ_{A1} and Φ_{A2}. This phase shift disrupts the intended timing, causing amplification to occur after sampling, which leads the ADC into a non-functional state. To address this vulnerability, the final flip-flop stage is equipped with a reset mechanism driven by either Φ_{S1} or $\Phi_{S1,R}$. The recovery behavior is illustrated by the blue waveform in Fig. 7.13b, which shows that the control signals return to normal within one sampling period after an SEU event. For the first two divider stages, however, no appropriate preceding signals are available to serve as a reset. As a result, TMR is adopted to reinforce fault tolerance against SEEs. With this mitigation in place, the entire clock generation circuit achieves robustness against single-event disruptions caused by heavy-ion impacts (Fig. 7.15).

7.4.3 Comparator and Time-Out Protection

The proposed ADC employs asynchronous logic to reduce overall conversion time and eliminate the need for a high-frequency sampling clock. In such systems, the comparator's decision time increases as the differential input voltage decreases. When the input signals

7.4 Design Details of the Key ADC Sub-blocks

Fig. 7.15 Simple D-FF is hardened by either TMR or additional reset

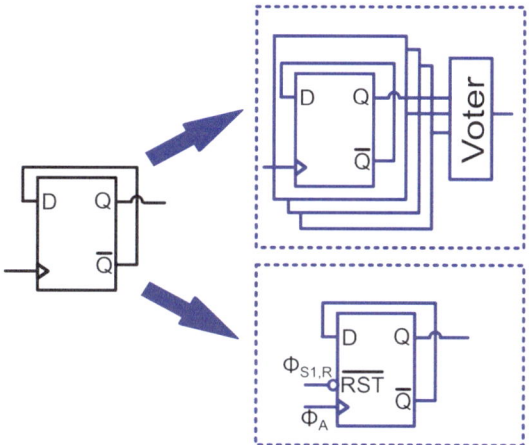

are nearly equal, the comparator may require an extended time to resolve the decision, potentially leading to incomplete comparisons for subsequent bits [211]. Although thermal noise has been used in previous work [159] to limit the comparison time, this approach is typically suited to low-speed, single-stage SAR ADCs and is not applicable for high-speed or multi-stage designs. Therefore, incorporating time-out protection becomes essential in asynchronous SAR logic.

A reference pulse with duration t_{TO} is introduced to define the maximum allowable comparison time. If the comparator fails to generate a completion signal Φ_{FIN} within this window, the SAR logic proceeds and assigns a predetermined value to the current bit. Previous implementations have generated t_{TO} using inverter chains [212] or the propagation delay of flip-flops [213, 214]. However, such fixed-delay approaches are not well-aligned with the comparator's behavior across process, voltage, and temperature (PVT) variations. To ensure proper operation in all corners, t_{TO} must exceed the worst-case comparison delay. Consequently, the time-out interval may become unnecessarily long to pass all corners, resulting in degraded comparator time efficiency.

The architecture of the proposed comparator with integrated time-out protection is depicted in Fig. 7.16a. This comparator consists of two main components: the primary comparison unit and the time-out protection block. Unlike conventional designs that rely on inverter chains or RC delay circuits, the time-out interval t_{TO} is generated using a dedicated dummy comparator. This dummy unit mirrors the architecture of the main comparator but is dimensionally scaled down to one-quarter size. This downsizing reduces its speed and power consumption while still tracking the PVT behavior relative to the main comparator. The dummy comparator inputs are connected to V_{refP} and V_{refN}. Through proper scaling, t_{TO} consistently exceeds $t_{0.5LSB}$, which is the time required for the main comparator to resolve a 0.5 LSB differential input. Post-layout simulations confirm that t_{TO} remains approximately $t_{0.5LSB} + 120$ ps across PVT corners. To enhance robustness against SEEs,

Fig. 7.16 Proposed comparator with time-out protection and SET hardening **a** block diagram, **b** timing diagram, **c** unit comparator schematic

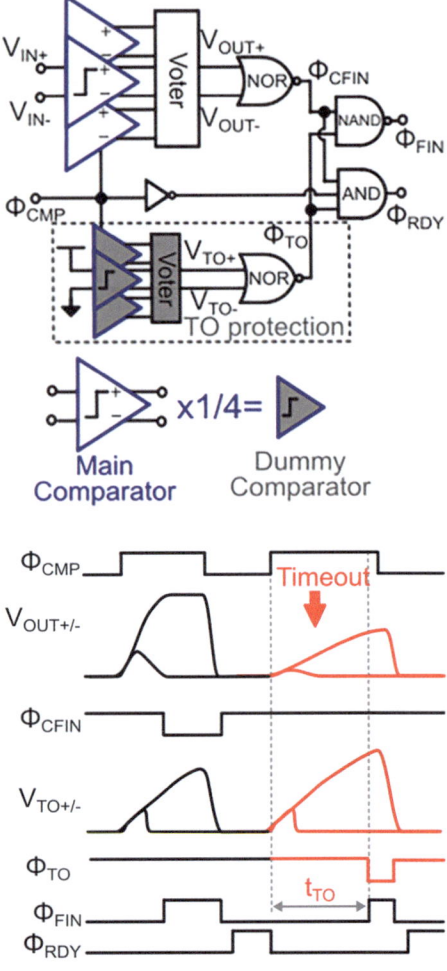

Fig. 7.17 Proposed comparator with time-out protection and SET hardening **a** block diagram, **b** timing diagram, **c** unit comparator schematic

both the main and dummy comparators are partitioned into three identical unit blocks. This segmentation increases SEE tolerance without impacting power consumption, noise performance, or offset, as the overall comparator area remains unchanged. The timing behavior of the proposed comparator is shown in Fig. 7.17b. Both comparison and timeout tracking begin and reset simultaneously, governed by the control signal Φ_{CMP}. The falling edge of Φ_{CMP} is triggered by Φ_{FIN}, which asserts high when either the main comparator finish signal Φ_{CFIN} or the time-out signal Φ_{TO}. The detailed transistor-level schematic of the unit comparator is shown in Fig. 7.18. It comprises a two-stage pre-amplifiers followed by a latch-based output stage, achieving fast response, minimal kickback noise, and low input offset. Moreover, the comparator is controlled using a single control signal, avoiding differential control paths and thus minimizing timing mismatches [215].

7.4 Design Details of the Key ADC Sub-blocks

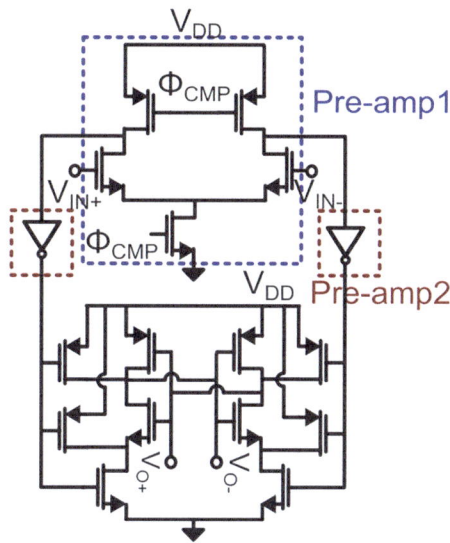

Fig. 7.18 Unit comparator schematic (dummy comparator has the same structure)

7.4.4 Digital Circuits and Transistors

In this prototype, several additional strategies are employed to further improve radiation robustness:

- All digital logic is implemented using a silicon-verified radiation-hardened standard cell library, in which both logic gates and flip-flops are hardened at the architectural and device levels. For instance, flip-flops adopt the DICE configuration to improve resilience against SEUs [216]. The corresponding layout is optimized to minimize the likelihood of double-node upsets caused by a single ion strike.
- In the analog domain, all NMOS transistors are enclosed within Deep N-well regions to reduce the amount of charge collected during a SEE [72].
- Guard rings are strategically placed between PMOS and NMOS active areas to mitigate the risk of SEL [217].
- Minimum-length transistors are avoided in the design to limit the impact of TID effects, which are more pronounced in narrow and short-channel devices [50].

7.4.5 Layout of the Coarse Stage ADC

The layout is also very important for the final ADC linearity, as the identical components may experience different processes and mismatches. The CDAC of the coarse ADC stage is the

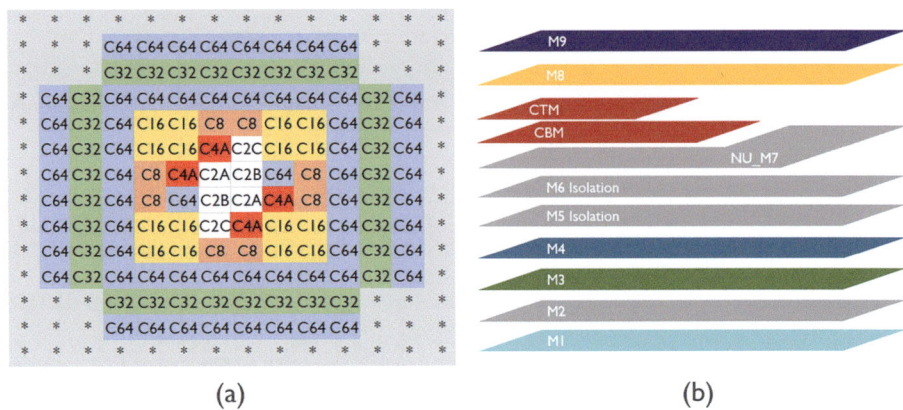

Fig. 7.19 Coarse stage CDAC: **a** common-centroid capacitor matrix, **b** unit MIM capacitor with isolation

most critical as it dominates the first 6 MSBs. Therefore, a common-centroid layout structure is used to improve the matching between the metal-insulator-metal (MIM) capacitor CDAC, as shown in Fig. 7.19a. By symmetrically arranging the unit capacitors around a common centroid, this approach minimizes fluctuations in parasitic effects, resulting in improved device matching and ADC linearity while reducing errors. In addition, it helps to reduce the process variations that occur during capacitor fabrication, which improves the reliability of the CDAC.

In addition to capacitor matching, metal connections introduce extra capacitance into the CDAC. The metal connections below the MIM capacitors are very crucial, which can introduce additional parasitic capacitance between the top and bottom plates of the MIM capacitor and lead to a non-binary capacitance ratio in the CDAC [218]. As a result, the linearity may drop significantly because of this. In the layout, two metals (metal 5 and 6) are connected to the ground and are used as isolation layers to shield the coupling between the metal connections (metal 3 and 4) and the MIM capacitors, as shown in Fig. 7.19b. In this way, the parasitics caused by the metal interconnect can be effectively reduced.

Figure 7.20a shows the positive side of the coarse stage ADC, which consists of a CDAC, a mismatch calibration block, an input bootstrap switch, reference switches, SAR logic and a comparator. The common node (all top plates of the capacitors) of the CDAC is connected to the mismatch calibration block, the comparator and the common mode reference switch (green lines in the figure). The bottom plates of the CDAC are connected to the reference switches and also to the input bootstrap switch (yellow lines in the figure). The detailed connections of the bottom and top plates are shown in Figs. 7.20b, c, respectively. Note that the metal connections on the bottom plate shown in Fig. 7.20b are well isolated by metals 5 and 6, effectively minimizing the nonlinear capacitance. The calibration block can be seen in Fig. 7.20d, which contains calibration switches and calibration capacitors. The calibration

7.5 Experimental Results

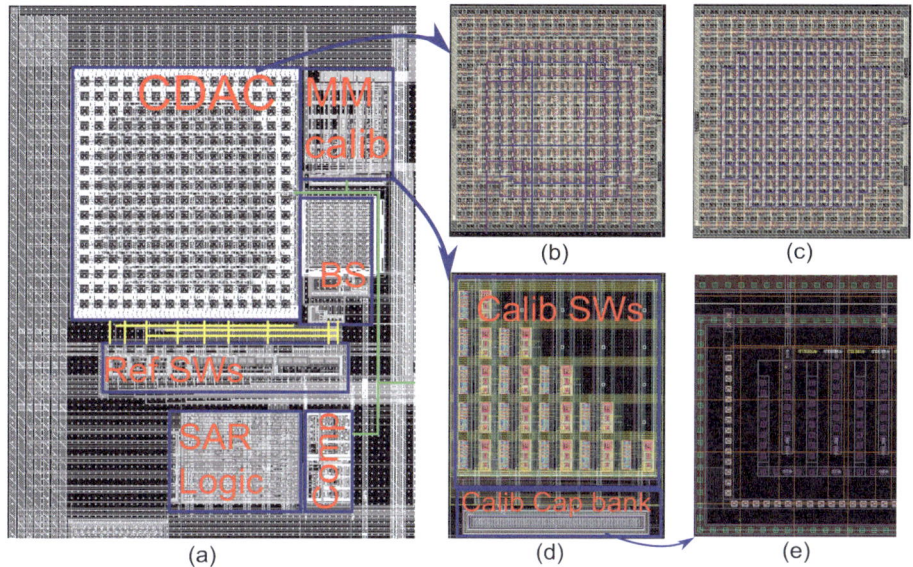

Fig. 7.20 Layout of **a** The positive side of a coarse stage ADC, **b** bottom plate connections of the CDAC, **c** top plate connections of the CDAC, **d** mismatch calibration block, **e** MOM capacitor bank for mismatch calibration

capacitor bank consists of single-finger metal-oxide-metal (MOM) capacitors arranged in a common-centroid layout, as shown in Fig. 7.20e. The mismatch calibration is performed by adding or subtracting the MOM capacitors from the CDAC as described in Chap. 6.

7.5 Experimental Results

The proposed ADC was fabricated using a 65 nm CMOS process, and the top-level layout of the prototype ADC die is shown in Fig. 7.21 and the ADC core locates in the center. Figure 7.22 demonstrates the details of the ADC core. The active area of the ADC core measures 778 µm × 1291 µm. As illustrated, the layout of the two channels is horizontally symmetric, with the RA centrally placed between the C-ADC and F-ADC. The clock generation circuitry is positioned at the center of the ADC to ensure balanced clock distribution to all sub-ADCs and the RA.

The performance evaluation of the ADC includes both standard electrical measurements and radiation testing. Initial characterization was conducted without radiation exposure to assess static and dynamic behavior. Calibration of offset, RA gain error, and bit weights was performed off-chip. For radiation testing, pulsed laser irradiation was used to evaluate SEE sensitivity via Two-Photon Absorption (TPA) laser, while TID sensitivity was assessed using X-ray exposure. These radiation experiments were conducted at the RELY laboratory, KU Leuven, in Geel, Belgium.

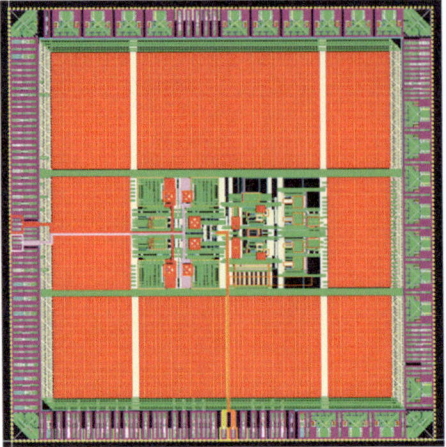

Fig. 7.21 Prototype ADC die photo

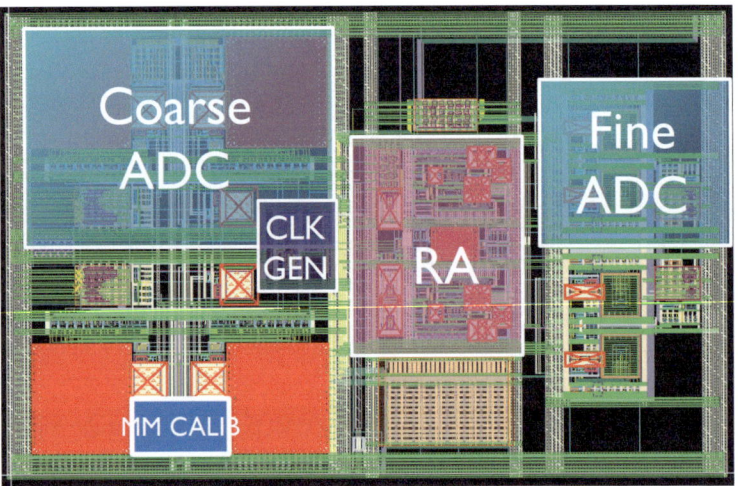

Fig. 7.22 Prototype ADC die photo

7.5.1 Electronic Measurement Results

The static performance of the prototype ADC was evaluated using the histogram method with a 10 MHz sinusoidal input signal sampled at 80 MS/s. As shown in Fig. 7.23, the DNL remains within ±0.7 LSB, and the INL stays within ±1.38 LSB. For dynamic performance characterization, sinusoidal inputs with amplitudes ranging from −0.3 to −1 dBFS were applied. At a sampling rate of 80 MS/s and an input frequency of 38 MHz, frequency-domain analysis of the ADC output using the DFT revealed an SNDR of 70.8 dB and an SFDR of

7.5 Experimental Results

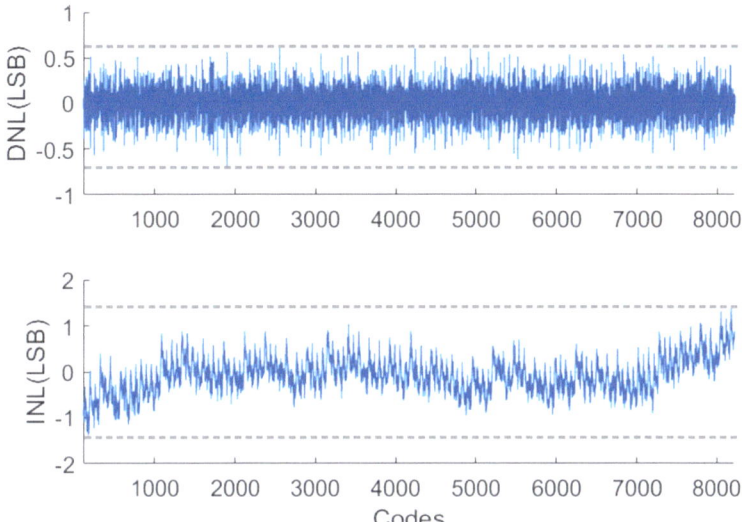

Fig. 7.23 Measured DNL and INL

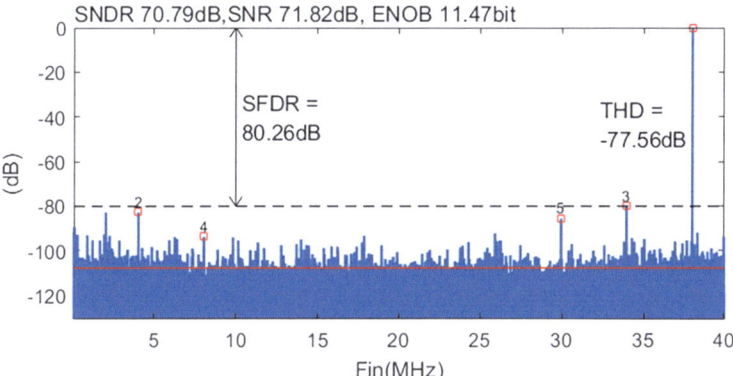

Fig. 7.24 Measured spectrum at $f_s = 80$ MS/s and $f_{in} = 38$ MHz

80.3 dB. The corresponding output spectrum is presented in Fig. 7.24. Under these Nyquist input conditions, the ADC consumes 13.8 mW of power from a 1.2 V supply. The overall metrics are summarized in Table 7.1.

A comprehensive analysis of the system's power distribution is illustrated in the pie chart of Fig. 7.25. In the case of the RA, parameters were configured as $M = 15 \times 32$ (number of AZ cycles) and $N = 16 \times 1$ (number of amplification cycles), aiming to optimize energy efficiency. Under these settings, the RA's power consumption equates to $1.016 \times P_{op-amp}$, which accounts for approximately 25% of the total power usage. Among all blocks, the digital section registers the highest power consumption (42%), primarily due to the implementation

Table 7.1 Performance summary of the prototype ADC

Parameter	Value
Technology	65 nm CMOS
Supply voltage	1.2 V
Core area	778 μm × 1291 μm
Sampling rate	80 MS/s
Input frequency (for dynamic test)	38 MHz
Input amplitude	-0.3 to -1 dBFS
DNL (max)	±0.7 LSB
INL (max)	±1.38 LSB
SNDR	70.8 dB
SFDR	80.3 dB
Power consumption	13.8 mW
Time-out protected comparator	Yes
Radiation-hardening techniques	Yes (SEE and TID)

of radiation-hardened gates and flip-flops designed to withstand SET. However, quantifying the exact power overhead from SEE mitigation is challenging, as hardening techniques are applied across multiple design levels—including the transistor, circuit, architectural, and layout levels. For instance, deploying DICE flip-flops can increase power usage by at least 50% compared to standard designs [80, 219]. Besides higher power draw, DICE structures introduce additional signal delays, which require delay compensation for certain digital paths, further contributing to power consumption. Additionally, constraints on transistor sizing in logic gates—implemented to enhance TID resilience—lead to increased power usage and latency.

The dynamic performance of the system was also assessed at varying sampling rates f_s and input frequencies f_{in}, while maintaining a constant input amplitude. Figure 7.26a presents results for $f_s = 80$ MS/s across different f_{in} values, demonstrating that the prototype achieves SNDR greater than 69 dB and SFDR exceeding 75 dB across the measured input frequencies. Figure 7.26b explores performance over a sampling rate range from 50 to 100 MS/s at a fixed input frequency $f_{in} = 20$ MHz. Results indicate that the ADC maintains stable performance up to 80 MS/s.

7.5.2 Irradiation Test Results

Laser and X-ray radiation evaluations were conducted to assess the prototype ADC's tolerance to SEEs and TID effects. The experimental setups for both X-ray and Laser tests are

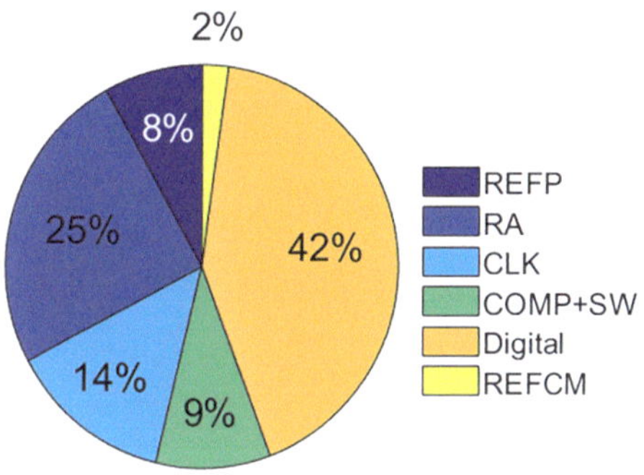

7.5 Experimental Results

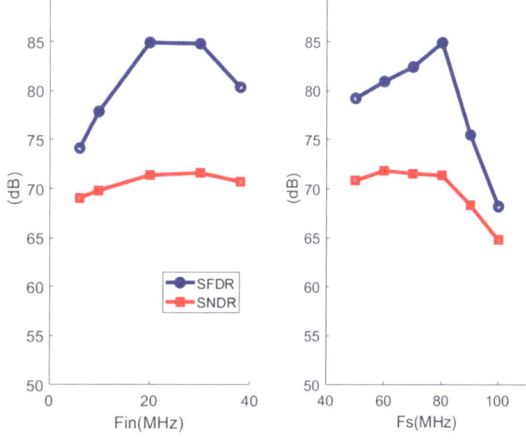

Fig. 7.25 Power breakdown at $f_s = 80$ MS/s and $f_{in} = 38$ MHz

Fig. 7.26 Measured SNDR and SFDR at **a** $f_s = 80$ MS/s versus f_{in}, **b** $f_{in} = 20$ MHz input versus f_s

shown in Figs. 7.27 and 7.29. In the setup, the ADC die was directly mounted on the PCB, with a circular opening of 1 mm radius on the backside to expose the die for irradiation. For the X-ray tests, two samples were used, while an additional two were assigned for the Laser tests. One reference sample, which was not subjected to any irradiation, served as a baseline. All measurement conditions were consistent with those used during standard electronic characterization.

7.5.2.1 Laser Test

The TPA laser test is an effective approach to emulate the impact of heavy-ion interactions in silicon devices [220, 221]. In a TPA test, a laser generates two photons, each with energy below the silicon bandgap but above half its value. When these photons are absorbed simultaneously, their combined energy is sufficient to excite electrons into the conduction band, thereby inducing transient currents in the silicon substrate [221]. Because TPA requires two photons to interact with the same electron nearly simultaneously, absorption only occurs at locations with high photon density—specifically near the focal point of the laser beam.

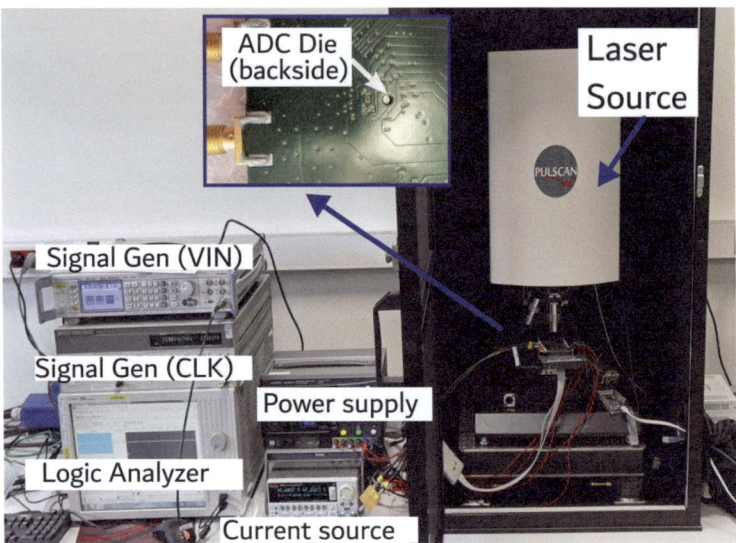

Fig. 7.27 Test setup for **a** laser test, **b** X-ray test

This property allows for precise spatial scanning of the device under test in the x, y, and z directions by adjusting the focus of the laser. In the conducted laser evaluation, a PULSBOX 2P system from PULSCAN was employed for the TPA experiments [222]. Laser energy was varied between 0.5 and 1.4 nJ, using 10 ns pulse widths at a repetition rate of 10 kHz. The active region of the ADC die was scanned using laser spot sizes of 20, 5, and 1 μm. During the test, the ADC operated at 40 MS/s with a DC input signal. An SEE event was defined as an output deviation exceeding ± 5 LSBs. Notably, SEE events were only detected with the 1 μm spot size, which provides the highest photon density and induces more current injection at equivalent laser energy.

Figure 7.28a, b illustrate the threshold energy required to induce SETs and the associated recovery time at the 1 μm spot size, respectively. Only half of the chip is displayed due to the nearly symmetric layout of the dual-channel architecture. Within both C-ADC and F-ADC blocks, the comparators demonstrated full immunity to laser-induced current. However, SEUs were observed in the switches, SAR logic, and comparator voting circuits, though these were recovered within a single sample period. In the clock generation circuitry, only single-sample SEUs were noted, and no SEFI incidents occurred throughout the test duration. The op-amp embedded in the RA block showed strong resilience to SEEs, with SEUs detected only in the NMOS input pair and NMOS gain-boosting amplifiers. No SEE effects were registered in the class-AB output stages, PMOS gain-boosting amplifiers, common-mode

7.5 Experimental Results

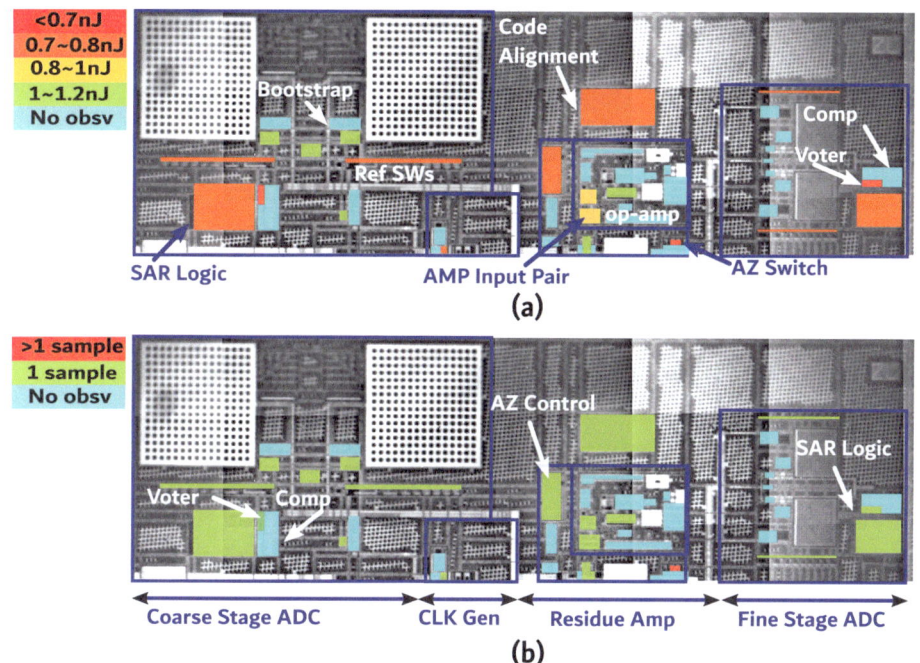

Fig. 7.28 Laser measurement results of **a** threshold laser energy for erroneous ADC outputs, **b** recovery time after erroneous outputs

feedback circuitry, or the cascaded stages and their biasing networks. However, the AZ switches within the RA exhibited low energy thresholds for SEE activation and, in rare instances, multi-sample output errors. These occurred only at 1.2 nJ laser energy due to the small geometries and interleaved layout of the switches, which allows partial charge sharing of the injected SEE current. Throughout the entire TPA test campaign, no SEL events were observed.

7.5.2.2 X-Ray Test

In space applications, backup ADCs are typically included alongside the main ADC to mitigate risks of functional degradation due to TID exposure [223]. If the main ADC becomes non-operational, a rapid switch-over mechanism activates one of the spare ADCs. As a result, an ADC may operate under three distinct conditions: Conversion mode (powered on with active clock and input signals), hot spare mode (powered on without clock or input), and cold spare mode (completely powered off). According to [224], the most severe TID-induced

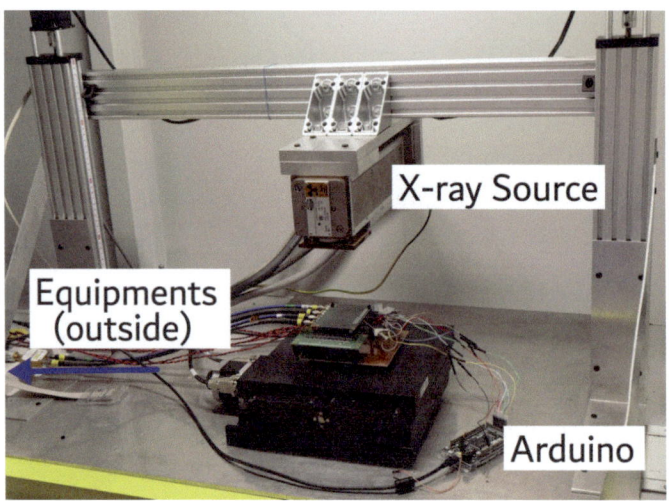

Fig. 7.29 Test setup for **a** laser test, **b** X-ray test

Fig. 7.30 Measured SNDR and SFDR when the ADC works in conversion and hot spare modes versus radiation dose

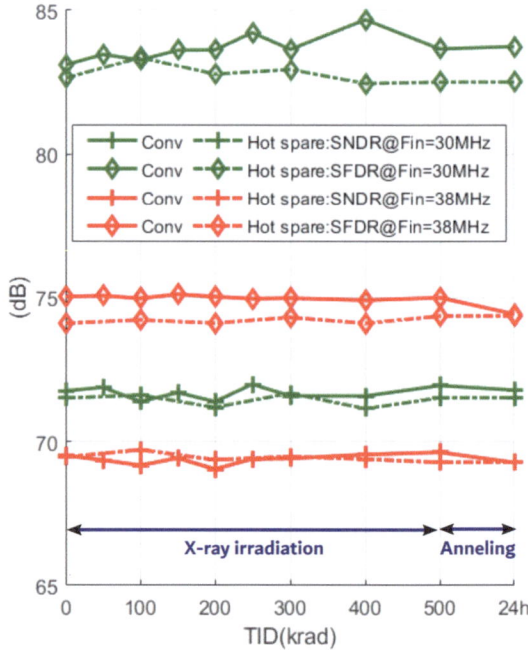

degradation tends to occur when an electric field is present across the transistors, implying that only power-on states are of primary concern. Therefore, the TID evaluation focused exclusively on the conversion and hot spare modes. For the conversion mode TID test, the ADC was operated with a sampling rate $f_s = 80$ MS/s and driven by a 30 MHz sine-wave input. In the hot spare configuration, supply and reference voltages were applied, while both clock and input signals were held at the common-mode level. Following irradiation, dynamic performance was assessed using $f_s = 80$ MS/s with input frequencies $f_{in} = 30$ MHz and 38 MHz.

A long fine-focus X-ray tube (PW 2240/20) with a anode (W) anode served as the radiation source, generating broadband X-ray exposure. The dose rate was set to 100 krad(Si) per hour, culminating in a total dose of 500 krad(Si). Performance measurements were conducted every hour, corresponding to every 100 krad increment, and compared against pre-irradiation benchmarks. Upon reaching the full dose, the devices underwent room-temperature annealing for 24 h, after which final measurements were recorded. The TID response is summarized in Fig. 7.30. Results from both operating modes reveal stable SNDR and SFDR across the full dose range, with performance remaining consistent even after annealing. In addition, overall power consumption showed no significant change before and after irradiation. These stable outcomes are consistent with expectations. As CMOS technology nodes scale down to 65 nm, the thinner gate oxide and reduced supply voltage help minimize charge trapping and mitigate threshold voltage shifts [57, 58]. Moreover, avoiding the use of minimum-sized transistors helps alleviate TID-related vulnerabilities at the transistor edges.

7.6 Performance Summarize

The prototype ADC performance is summarized in Table 7.2 and compared with state-of-the-art ADCs. The comparison includes both radiation-tolerant ADCs and regular ADCs. In general, radiation-tolerant ADCs have to spend more power and effort to achieve a similar accuracy and speed as the regular ADCs. Thanks to the Semi-Time-interleaved architecture, this work shows the best Walden FoM among the radiation-tolerant ADCs. Besides, this prototype also presents a comparable FoM to the non-radiation-tolerant ADCs.

7.7 Conclusion

This chapter introduces a 13-bit radiation-tolerant ADC realized in 65 nm CMOS technology. To enhance energy efficiency, a Semi-Time-Interleaved pipelined-SAR structure is employed, which brings both power efficiency and SEE tolerance. Each sub-block of the

Table 7.2 ADC performance comparison

	This work	ISSCC 2017 [205]	JINST 2013 [206]	VLSI 2020 [225]	VLSI 2019 [226]	ASSCC 2017 [227]
Resolution [bits]	13	14	12	12	12	13
Sampling frequency [MS/s]	80	75	40	12	200	160
Technology [nm]	65	65	130	40	40	65
Architecture	**Semi-TI Pipelined-SAR**	Pipelined-SAR	Pipeline	SAR	Pipelined-SAR	Pipelined-SAR
SNDR@Nyquist [dB]	70.8	70.8	67	59.6	62.1	61.5
SFDR@Nyquist [dB]	80.3	89.6	77.8	75.1	67.1	71.4
Power [mW]	13.8	24.9	54.5	0.472	3.9	22.1
FoMw [fJ/conv step]	60.7	117.2	685.4	50.7	19	162.6
FoMs [dB]	165.4	162.6	152.6	160.6	166.2	157.1
Radiation-tolerant	Yes	Yes	Yes	No	No	No

proposed ADC is radiation hardened at architectural, circuit, and layout levels for TID and SEE. The prototype ADC reaches 70.79-dB SNDR and 80.26-dB SFDR at the Nyquist input frequency under an 80 MS/s sampling rate. Moreover, TID irradiation tests confirm that the ADC remains unaffected up to 500 krad(Si), displaying robust resilience. Notably, the ADC exhibits a limited SEE-sensitive region and swift recovery even in the occurrence of SEE events.

Conclusions and Future Research

Abstract

This chapter provides a concise summary of the key steps undertaken throughout the book to achieve the overarching objective: the development of a high-performance, radiation-tolerant analog-to-digital converter (ADC). It traces the design flow from initial architectural exploration to radiation-aware circuit implementation and verification. Special emphasis is placed on the research contributions, including a novel hardening methodology that integrates layout-level techniques with system-level redundancy, and an innovative calibration scheme that enhances robustness under radiation-induced degradation. These advancements distinguish the proposed ADC from conventional designs in both performance and resilience. Finally, the chapter outlines potential directions for future research, such as reducing power and area overhead, implementing adaptive radiation monitoring, and extending the design framework to support emerging process technologies. These suggestions aim to inspire continued progress in the field of radiation-hardened mixed-signal design.

8.1 Introduction

This chapter briefly summarizes the steps taken to achieve the ultimate goal of developing a high-performance, radiation-tolerant ADC. It then highlights the research novelties and provides some suggestions for future research.

8.2 General Conclusions

This work includes a comprehensive investigation into the development of high-performance, radiation-tolerant ADCs for space applications. The challenges in developing conventional and radiation-tolerant ADCs are highlighted in Chap. 1, which outlines the research objectives. Chapter 2 lays the foundations of radiation knowledge by examining the physical mechanisms behind radiation effects, focusing on TID effects and SEEs. In recent years, emphasis has been placed on SEEs, and techniques to mitigate these effects are discussed. Chapter 3 details the process of evaluating CMOS technology against radiation effects, leading to the formulation of a radiation-hardened ASIC/IC design flow. Chapter 4 evaluates the suitability of 65 nm CMOS technology for space projects, focusing on heavy ion testing to characterize ionization charge and pulse duration. Chapter 5 addresses the ADC architecture and performance trade-offs that are critical to the subsequent system-level ADC design. Chapter 6 analyzes a pipelined-SAR ADC system, optimizing power efficiency and addressing mismatch effects. Finally, Chap. 7 presents a radiation-tolerant 13-bit ADC prototype that demonstrates exceptional performance metrics, robustness to TID effects, and fast recovery from SEE events, thus meeting the research objectives.

8.3 Research Novelties

- The process of performance evaluation of CMOS technology under radiation is presented, which serves as a fundamental cornerstone for the subsequent evaluation of the radiation-protected design. A comprehensive radiation-hardened ASIC/IC design flow is also presented, which builds on the conventional design process and includes additional steps to account for radiation effects. All checks and steps ensure both electrical performance and radiation tolerance. This process provides a general guideline for the development of radiation-hardened ASIC/IC.
- A 65 nm test chip for characterizing the SET ionization charge and pulse duration was developed and tested under a heavy ion beam. Detailed measurement data were obtained from the heavy ion tests, which are very valuable to obtain an accurate SET current model for radiation-hardened IC performance prediction. This current model was used in the RHBD process in ADC. Most importantly, this model can also provide an accurate SEE evaluation when 65 nm technology is used for other designs.
- The design trade-off for a radiation-tolerant ADC is presented for the first time to reveal the relationship between power efficiency as well as radiation tolerance. This trade-off provides fundamental guidelines for radiation-tolerant ADC design at both the system and block levels.
- Several novel design techniques have been introduced to improve both the power efficiency and the radiation tolerance of an ADC. A semi-time-interleaved pipelined-SAR structure was invented to improve the usability of the residue amplifier and the overall power efficiency. A residue amplifier with ping-pong auto-zeroing was used to reduce power consumption further. A mismatch calibration technique is introduced to minimize the size of the CDAC and improve the linearity of the ADC and power consumption dur-

ing switching. Multiple biasing has been implemented to increase the SEE tolerance of the residue amplifier. A fully SEE-hardened clock generation is used to avoid complete ADC corruption due to SEEs. With the enhancements from both system and block levels, the prototype ADC achieves a total power consumption of 13.8 mW and a state-of-the-art Walden Figure of Merit of 60.7 fJ/conv step, yielding a comparable efficiency as the non-radiation-tolerant ADCs with similar specifications.

8.4 Future Research

In this research, a test chip Godzilla in 65 nm CMOS technology was developed and the technology used in the following ADC was evaluated for the first time. This chip, together with the heavy ion measurement, provides enormously valuable data for understanding the technology. A high-performance, radiation-tolerant ADC chip was then developed and tested according to the radiation-hardened IC flow. However, some improvements and further investigations can still be made on both chips. In addition, more advanced ADC designs can also be initiated based on this ADC prototype.

8.4.1 Further Investigation of 65 nm CMOS Technology

- **65 nm SET Current Striker Update**
  SET current modeling is a very effective tool to support RHBD as described in Chap. 2. The proposed ADC described in Chap. 7 was implemented using this SET current model to evaluate the SEE performance. This model is referred to as SET current striker in IMEC's DARE65 platform. However, this SET striker model was created based on a model from a 180 nm process with a technology scaling factor. Therefore, the model may either overestimate or underestimate the SEEs in 65 nm technology. The model can be updated by re-running the post-layout simulation with the LET values matching the LET value in the heavy ion test and comparing the difference in simulated SET charge/duration with that from the measurement.
- **65 nm SET Evaluation under Two-photon Absorption**
  The test vehicle for the 65 nm CMOS technology has shown the SET charge and duration performance of single transistors under heavy ion radiation. However, the heavy ion test is expensive and difficult to access (limited test facilities). In addition, troubleshooting the cause of the unexpected heavy ion results can be difficult because the size of the heavy ion beam is usually much larger than the size of the chip and the malfunction can affect any part of the chip. As a result, troubleshooting can be complicated and time-consuming as some phenomena are a product of the SEE and the specific state of the circuits. The

TPA laser experiment is a more cost-effective and flexible option for SEE assessment and troubleshooting [220, 221]. The spot size of the laser can reach several micrometers, which is much smaller than most die sizes. By tuning the focus of the laser, the TPA experiment enables precise scanning in the x, y and z axes of the target device. However, in the context of 65 nm technology, the correlation between the laser energy and the LET of the heavy ions has yet to be discovered. Therefore, the TPA measurement results cannot be directly related to the SEE performance. The extraction of the SET effect under the TPA laser energy is imperative to expand the SEE assessment capabilities.

To determine the relationship between laser energy and SET duration as well as pulse duration, a TPA laser test can be applied to the Godzilla chip. This experiment aims to find out the relationship between the laser energy and the SET characteristics. Subsequently, the correlation between laser energy and heavy ion LET will provide important insights into the SEE performance. This comprehensive information will provide an additional method to evaluate the SEE performance, complementing the heavy ion test and thus increasing the flexibility in evaluating radiation-tolerant designs.

8.4.2 Improvement of the Prototype ADC

- **CDAC and Mismatch Calibration Update of the Prototype ADC**
  The CDAC capacitor mismatch calibration from Chap. 6 was also implemented for the ADC described in Chap. 7. However, this calibration method cannot completely cover the deviation of the capacitor from the standard capacitor value, which leads to a certain loss of linearity in the measurement. One of the possible reasons for this deviation is the parasitic capacitance between the CDAC output and MOM capacitor bottom plate. Ideally, these parasitic capacitors should maintain a ratio of two between the two adjacent bits. But the connecting wires have degraded this ratio and caused poorer linearity. Therefore, the CDAC layout needs to be redrawn for better system-level matching. In addition, the calibration range should be enlarged to cover more error capacitance.
- **Heavy Ion Test for the Prototype ADC**
  The radiation-tolerant ADC was only evaluated with the TPA laser test. However, the heavy ion beam is closer to the real radiation environment in space and allows us to measure SEU cross-sections for the complete ADC. Therefore, it may be useful to measure the ADC error amplitude/rate and cross-section under a heavy ion beam with different LET.

8.4.3 Future ADC Designs

- **High-Performance Pipelined-SAR ADCs for Applications in Extreme Conditions**
  The electrical specification of the ADC prototype is suitable for lots of applications where the analog input signal needs to be digitized. These applications can also include other extreme working conditions, such as extremely low temperatures for quantum computers and high temperatures in a jet engine. With appropriate modifications, the ADC can also be used for these applications.
- **Ultra High Speed or Resolution ADCs for Space Applications**
  The structure of the ADC prototype has been proven to have excellent power efficiency and radiation tolerance. Therefore, higher specifications can be pursued based on this structure. For example, a higher sampling frequency can be achieved by making multiple channels work together with the time-interleaving method at the system level. Since each channel operates at a sampling rate of 80 MS/s, N time-interleaved structures can achieve a sampling rate of Nx80 MS/s. However, channel mismatch and clock skew dominate the performance degradation. Therefore, calibration techniques must be used to solve these problems. Moreover, applying such an ADC prototype to a more advanced technology, such as 22 nm FD-SOI or even 7 nm FinFET technology, can also further improve the specifications of the ADC, as advanced technologies have less parasitic elements and lower supply voltage. With the time interleaving technique and advanced technologies, the ADC can achieve ultra-high sampling speed, e.g. giga sample points per second. However, their radiation tolerance needs further investigation as they may have worse SEE tolerance as explained in Chap. 2. Another direction is to extend the prototype ADC to achieve a higher resolution. This can be achieved by using a lower sampling frequency, adding more bits to the CDAC or even by implementing the noise shaping technique in the coarse or fine stage of the ADC. From the design tradeoff presented in Chap. 5, it is clear that higher resolution leads to poorer SEE tolerance at the system level. Therefore, additional effort is required for hardening.

9. Research Valorization Feasibility Study

Abstract

This chapter explores the concept and process of research valorization, emphasizing its critical role in bridging the gap between academic research and real-world application. The goal of research valorization is to study the possibility of transforming academic achievements from research into practical applications and tangible solutions with social and economic impact. Through technology transfer, commercialization, and community engagement, valorization ensures that the valuable findings from doctoral research find meaningful applications in the real world. This process not only enhances the visibility and relevance of academic endeavors but also contributes to economic growth, innovation, and the overall betterment of society.

9.1 Introduction

Analog-to-Digital Converters (ADCs) are one of the most critical components in an electronic system. The physical world consists of analog information, but modern data processing can only deal with 0 and 1. Therefore, the ADC acts as a bridge between the analog and digital worlds. For example, an ADC is needed to convert an analog signal, such as a sound picked up by a microphone or light entering a digital camera, into a digital signal for further processing, such as recording and editing.

In addition to the consumer electronics market (such as cell phones, cameras, computers, etc.), ADCs are also used in aerospace, nuclear science, quantum computing, and biomedical fields. These applications require more critical working conditions than consumer applications. In space, the heavy ions can cause chips like ADCs to generate erroneous outputs, which can have serious consequences such as satellite malfunction. However, there are only a few companies that develop/produce ADC chips for these applications under extreme con-

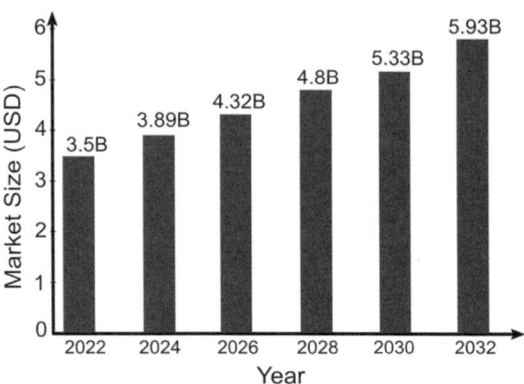

Fig. 9.1 Estimated global space semiconductor market size from 2022 to 2032 in USD (*Source* Precedence Research [228])

ditions. The most common method is to use COTS (Commercial-Off-The-Shelf) products and create some redundancies at the application level (PCB or software) rather than at the chip level. Nevertheless, these ADCs are designed for normal working conditions and consumer applications. They can reduce the cost and development time of the application. But they also reduce integrity and reliability and increase application-level design complexity.

In this research project, a prototype of a radiation-hardened high-performance ADC is developed and tested. The measurement results show that this chip has a high tolerance to SEEs and TID effects. In addition, the ADC is very power efficient and even shows better power efficiency than conventional ADCs. As mentioned above, this ADC will benefit space applications in terms of cost, area, flexibility, and reliability, which in turn is an advantage for research and industry (Fig. 9.1).

9.2 Present State of Space Market

The global space semiconductor market was valued at USD 3.5 billion in 2022 and is expected to reach USD 5.93 billion by 2032, at a compound annual growth rate (CAGR) of 5.42% during the forecast period from 2023 to 2032 [228]. In 2022, North America will play an important role with a share of more than 39% of sales. The Asia Pacific region is expected to exhibit the fastest CAGR during the forecast period. In terms of types, the radiation-hardened electronics segment dominated the market in terms of revenue share in 2022. In terms of components, integrated circuits will exhibit the highest CAGR during the forecast period. As for applications, satellites are expected to have the highest market share in 2022, while launch vehicles are expected to exhibit the fastest CAGR during the forecast period [228].

The global analog-to-digital converter market reached a value of USD 2.38 billion in 2021. However, the market is forecast to reach a value of USD 3.33 billion by 2027, at a CAGR of 5.40% during the period 2022–2027 [229]. The radiation-hardened data converter market

9.3 Offer to the Market

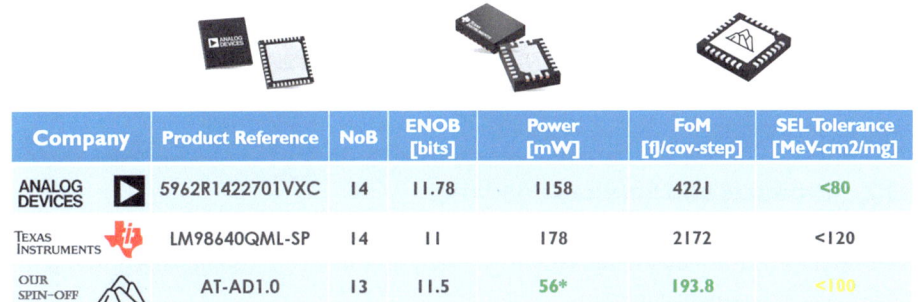

*The prototype ADC core consumes 14mW, auxiliary circuits are estimated to be 42mW based on previous silicon-verified design

Fig. 9.2 Comparison between AT-AD1.0 and other radiation-hardened ADC in market

size is also expected to develop revenue and exponential market growth at a remarkable CAGR during the forecast period from 2023 to 2030. The growth of the market can be attributed to the increasing demand for radiation-hardened data converters owing to the aerospace, national defense, and other applications on a global scale [230].

9.3 Offer to the Market

A spin-off can be a possible valorization plan to offer customized design services for projects and products that need to function under extreme conditions such as radiation and extremely low/high temperatures. The spin-off's first product is based on the prototype chip from this research, which is described in Chap. 7. After the spin-off has received the profits from the first product, it will expand the product lines and offer a higher-level solution.

9.3.1 AT-AD1.0 High Performance Radiation-Hardened ADC

The first product of the spin-off will be based on the radiation-hardened ADC described in Chap. 7, and the model number is defined as AT-AD1.0. However, the prototype from this thesis still needs some auxiliary blocks, such as bandgap voltage references and low-voltage differential signaling (LVDS), to be a complete product. Also, additional tests and improvements need to be done to increase the yield of a product.

Figure 9.2 compares the high-performance ADC AT-AD1.0 with commercial ADCs with similar specifications on the market. It can be seen that the AT-AD1.0 has similar linear accuracy (ENOB) but much lower power consumption and FoM. This means that the power efficiency of the AT-AD1.0 is 10x higher than that of the LM98640QML-SP from Texas Instruments and 20x higher than that of the 5962R1422701VXC from Analog Devices. In addition, the SEL tolerance is expected to be close to the other two products. Lower power consumption means that fewer power sources are needed in spacecraft, reducing the weight of solar cells and batteries as well as the complexity and risk of lunch.

9.3.2 Products and Services Roadmap

The overall plan for the development of products and services is summarized in Fig. 9.3, which is divided into three phases.

In the initial phase, development is mainly focused on the AT-AD1.0 high-performance radiation-hardened ADC. With the design of additional auxiliary circuits and tests, this ADC can be launched on the market within a year. After that, the AT-AD1.X ADC family can be developed based on the AT-AD1.0 ADC with modifications or additional hardening. These ADCs target different applications, such as low-power for consumer applications, low-temperature tolerance for cryogenic applications or high-speed for telecommunication purposes. The investment in the initial phase is the highest compared to the other phases due to the cost of renting the office, hiring employees, purchasing EDA licenses and servers, etc.

After the release of the AT-AD1.0 and later the ADC family, the spin-off is expected to benefit from the market. Then several analog/mixed-signal IP/chips for extreme conditions will be initiated to have various analog/mixed-signal products such as ADC, PLL and digital IPs. When these IPs/products enter the market, a stable and substantial profit will be realized, which can support the spin-off to enter phase three. By phase three, the spin-off has developed several products in the analog/mixed-signal domain. With a further expanded design team, some higher ASIC solutions and digital front-end and back-end designs can then be provided to customers.

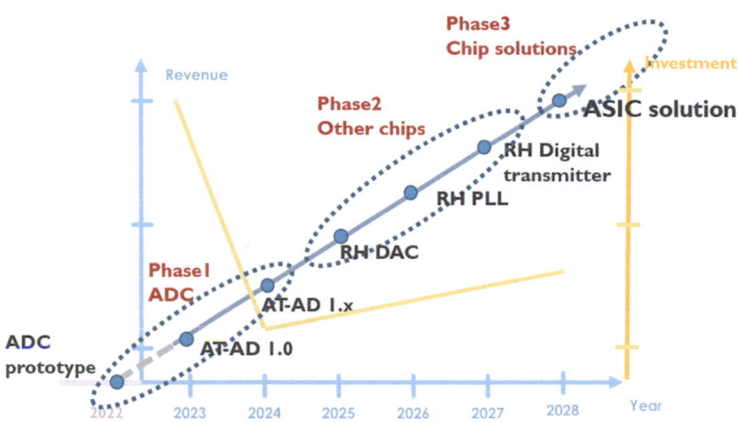

Fig. 9.3 Products and services roadmap of the spin-off

9.3.3 Business Model

The business model for generating profit for the spin-off can be summarized in three main categories:

- **Licensing/selling the design in white box**
  The first option is to sell or license the design/IP in the white box, i.e. the customer has access to the complete design details such as the detailed design structures, the dimensions of the components and the simulation test benches. This option offers the customer the greatest flexibility as they can modify the design to suit their needs. They can also flexibly integrate their own design and IP to achieve optimal performance and area efficiency. Of course, this option commands the highest price as all technical details are disclosed unconditionally.
- **Licensing/selling the design in black box**
  The design/IP can also be transmitted in the black box. In this way, the customer does not have access to the technical details, but can receive the final layout of the design/IP. This option offers a moderately flexible integration option for the customer, as they can still freely integrate the design/IP into their own design. However, they must retain the design and specifications. The cost of a black box is lower than that of a white box.
- **Selling the complete chips**
  The last business option is to sell the complete chips. These chips come from our design/IP with some additional circuitry. This option is the only one where the spin-off has to be responsible for the tapeout, packaging and yield. The customer receives a chip with packaging and a user manual that is ready for immediate use. This option is the cheapest compared to the other two options. In addition, this complete chip offers the highest reliability as the product is fully developed by the spin-off and we take into account all side effects due to manufacturing and packaging. However, the design flexibility for the customer is very limited as they can only make the design at the board level.

9.4 Conclusion

Valorization of research is a crucial link between academic achievements and practical applications and aims to translate research results into tangible solutions with social and economic impact. The valorization plan in this chapter foresees the creation of a spin-off company whose first product is based on the radiation-hardened high-performance ADC designed in Chap. 7. Subsequently, several analog/mixed-signal IPs will be developed for various applications and purposes. Finally, the spin-off is able to provide not only the chip design and manufacturing, but also the solution services. The ultimate vision of the spin-off is to build a comprehensive portfolio of analog/mixed-signal solutions and advanced ASIC design services, making the spin-off a major player for suppliers of microelectronic components in extreme environments.

References

1. A.-R. O. Bello, "A Comprehensive Guide to SpaceX Starlink Satellites," Apr 2023. [Online]. Available: https://interestingengineering.com/innovation/comprehensive-guide-to-spacex-starlink-satellites.
2. B. Dunbar, "What is Artemis," Jul 2019. [Online]. Available: https://www.nasa.gov/what-is-artemis.
3. M. J. Pelgrom and M. J. Pelgrom, *Analog-to-Digital Conversion*. Springer, 2013.
4. A. Colagrossi and M. Lavagna, "A Spacecraft Attitude Determination and Control Algorithm for Solar Arrays Pointing Leveraging Sun Angle and Angular Rates Measurements," *Algorithms*, vol. 15, no. 2, 2022.
5. Z. Zhao, G. Chen, and B. Wang, "Test Method for SADA's Servo Control System of China Space Station," in *2019 IEEE 10th International Conference on Mechanical and Aerospace Engineering (ICMAE)*, 2019, pp. 297–301.
6. L. Danial, N. Wainstein, S. Kraus, and S. Kvatinsky, "Breaking Through the Speed-Power-Accuracy Tradeoff in ADCs Using a Memristive Neuromorphic Architecture," *IEEE Transactions on Emerging Topics in Computational Intelligence*, vol. 2, no. 5, pp. 396–409, 2018.
7. M. Steyaert and K. Uyttenhove, "Speed-Power-Accuracy Trade-off in High-Speed Analog-to-Digital Converters: Now and in the Future..." *Analog Circuit Design: High-Speed Analog-to-Digital Converters; Mixed Signal Design; PLL's and Synthesizers*, pp. 3–24, 2000.
8. R. Walden, "Analog-to-Digital Converter Survey and Analysis," *IEEE Journal on Selected Areas in Communications*, vol. 17, no. 4, pp. 539–550, 1999.
9. A. M. A. Ali, A. Morgan, C. Dillon, G. Patterson, S. Puckett, P. Bhoraskar, H. Dinc, M. Hensley, R. Stop, S. Bardsley, D. Lattimore, J. Bray, C. Speir, and R. Sneed, "A 16-bit 250-MS/s IF Sampling Pipelined ADC With Background Calibration," *IEEE Journal of Solid-State Circuits*, vol. 45, no. 12, pp. 2602–2612, 2010.
10. M. J. Gadlage, R. D. Schrimpf, J. M. Benedetto, P. H. Eaton, D. G. Mavis, M. Sibley, K. Avery, and T. L. Turflinger, "Single Event Transient Pulse Widths in Digital Microcircuits," *IEEE Transactions on Nuclear Science*, vol. 51, no. 6, pp. 3285–3290, 2004.
11. V. Ferlet-Cavrois, L. W. Massengill, and P. Gouker, "Single Event Transients in Digital CMOS – A Review," *IEEE Transactions on Nuclear Science*, vol. 60, no. 3, pp. 1767–1790, 2013.
12. R. Baumann and K. Kruckmeyer, *Radiation Handbook for Electrics*. [Online]. Available: https://www.ti.com/seclit/eb/sgzy002a/sgzy002a.pdf.

13. B. Murmann, "ADC Performance Survey 1997-2023," [Online]. Available: https://github.com/bmurmann/ADC-survey.
14. H. Xu, Y. Cai, L. Du, Y. Zhou, B. Xu, D. Gong, J. Ye, and Y. Chiu, "28.6 A 78.5dB-SNDR Radiation- and Metastability-Tolerant Two-Step Split SAR ADC Operating up to 75MS/s with 24.9mW Power Consumption in 65nm CMOS," in *2017 IEEE International Solid-State Circuits Conference (ISSCC)*, 2017, pp. 476–477.
15. J. Kuppambatti, J. Ban, T. Andeen, P. Kinget, and G. Brooijmans, "A Radiation-Hard Dual Channel 4-Bit Pipeline for a 12-bit 40 MS/s ADC Prototype with Extended Dynamic Range for the ATLAS Liquid Argon Calorimeter Readout Electronics Upgrade at the CERN LHC," *Journal of Instrumentation*, vol. 8, no. 09, p. P09008, 2013.
16. Analog Devices, "AD9246S Datasheet," Nov 2017. [Online]. Available: https://www.analog.com/en/products/ad9246s.html.
17. Texas Instruments, "LM98640QML-SP Datasheet," Nov 2018. [Online]. Available: https://www.ti.com/lit/gpn/lm98640qml-sp.
18. A. P. French, *An Introduction to Quantum Physics*. Routledge, 2018.
19. D. J. Thomas, "ICRU Report 85: Fundamental Quantities and Units for Ionizing Radiation," 2012.
20. J. F. Ziegler, "The Stopping and Range of Ions in Matter homepage. SRIM software version 2013," [Online]. Available: http://www.srim.org/SRIM/SRIM2011.htm.
21. R. Baumann, "Investigation of the Effectiveness of Polyimide Films for the Stopping of Alpha Particles in Megabit Memory Devices," Texas Instruments Tech. Rep, Tech. Rep., 1991.
22. F. B. McLean and T. R. Oldham, "Basic Mechanisms of Radiation Effects in Electronic Materials and Devices," *Harry Diamond Laboratories Technical Report*, p. 2129, 1987.
23. J. R. Srour and J. W. Palko, "Displacement Damage Effects in Irradiated Semiconductor Devices," *IEEE Transactions on Nuclear Science*, vol. 60, no. 3, pp. 1740–1766, 2013.
24. L. Zhang, S. Bi, and M. Liu, "Lightweight Electromagnetic Interference Shielding Materials and Their Mechanisms," *Electromagnetic materials and devices*, pp. 1–10, 2018.
25. J. R. Schwank, M. R. Shaneyfelt, D. M. Fleetwood, J. A. Felix, P. E. Dodd, P. Paillet, and V. Ferlet-Cavrois, "Radiation Effects in MOS Oxides," *IEEE Transactions on Nuclear Science*, vol. 55, no. 4, pp. 1833–1853, 2008.
26. S. Duzellier, "Radiation Effects on Electronic Devices in Space," *Aerospace science and technology*, vol. 9, no. 1, pp. 93–99, 2005.
27. C. Claeys and E. Simoen, *Radiation Effects in Advanced Semiconductor Materials and Devices*. Springer Science & Business Media, 2002, vol. 57.
28. A. Lechner, "CERN: Particle Interactions with Matter," *CERN Yellow Rep. School Proc.*, vol. 5, p. 47, 2018.
29. M. G. Moore, "Three Types of Interaction," 1989.
30. R. N. Hamm, J. E. Turner, H. A. Wright, and R. H. Ritchie, "Heavy-Ion Track Structure in Silicon," *IEEE Transactions on Nuclear Science*, vol. 26, no. 6, pp. 4892–4895, 1979.
31. W. Stapor, P. McDonald, A. Knudson, A. Campbell, and B. Glagola, "Charge Collection in Silicon for Ions of Different Energy but Same Linear Energy Transfer (LET)," *IEEE Transactions on Nuclear Science*, vol. 35, no. 6, pp. 1585–1590, 1988.
32. Secretariat, ECSS, "ECSS-Q-HB-60-02A – Techniques for Radiation Effects Mitigation in ASICs and FPGAs Handbook," 2016.
33. G. Aad, X. S. Anduaga, S. Antonelli, M. Bendel, B. Breiler, F. Castrovillari, J. Civera, T. Del Prete, M. T. Dova, S. Duffin *et al.*, "The ATLAS Experiment at the CERN Large Hadron Collider," 2008.
34. L. Evans and P. Bryant, "LHC Machine," *Journal of instrumentation*, vol. 3, no. 08, p. S08001, 2008.

35. S. Wolf and R. Tauber, *Silicon Processing for the VLSI Era: Process Technology*, ser. Silicon Processing for the VLSI Era. Lattice Press, 2000.
36. J. Barth, C. Dyer, and E. Stassinopoulos, "Space, Atmospheric, and Terrestrial Radiation Environments," *IEEE Transactions on Nuclear Science*, vol. 50, no. 3, pp. 466–482, 2003.
37. K. S. Krane, *Introductory Nuclear Physics*. John Wiley & Sons, 1991.
38. J. F. Ziegler, "Terrestrial Cosmic Ray Intensities," *IBM Journal of Research and Development*, vol. 42, no. 1, pp. 117–140, 1998.
39. Survivability Vulnerability and Assessment Directorate (SVAD), "Test Operations Procedure (TOP) 1-2-612 Nuclear Environment Survivability," Tech. Rep., 2008.
40. J. A. Van Allen, C. E. McIlwain, and G. H. Ludwig, "Radiation Observations with Satellite 1958," *Journal of Geophysical Research*, vol. 64, no. 3, pp. 271–286, 1959.
41. E. Amato, "The Origin of Galactic Cosmic Rays," *International Journal of Modern Physics D*, vol. 23, no. 07, p. 1430013, 2014.
42. J. R. Höorandel, "A Review of Experimental Results at the Knee," vol. 47, no. 1, p. 41, 2006.
43. W. Apel, J. Arteaga-Velàzquez, K. Bekk, M. Bertaina, J. Blümer, H. Bozdog, I. Brancus, E. Cantoni, A. Chiavassa, F. Cossavella et al., "Ankle-Like Feature in the Energy Spectrum of Light Elements of Cosmic Rays Observed with KASCADE-Grande," *Physical Review D*, vol. 87, no. 8, p. 081101, 2013.
44. S. Plunkett and S. Wu, "Coronal Mass Ejections (CMEs) and Their Geoeffectiveness," *IEEE Transactions on Plasma Science*, vol. 28, no. 6, pp. 1807–1817, 2000.
45. T.-P. Ma and P. V. Dressendorfer, "Ionizing Radiation Effects in MOS Devices and Circuits," 1989.
46. S. Bala, R. Kumar, and A. Kumar, "Total Ionization Dose (TID) Effects on 2D MOS Devices," *Transactions on Electrical and Electronic Materials*, vol. 22, pp. 1–9, 2021.
47. J. R. Schwank, P. S. Winokur, F. W. Sexton, D. M. Fleetwood, J. H. Perry, P. V. Dressendorfer, D. T. Sanders, and D. C. Turpin, "Radiation-Induced Interface-State Generation in MOS Devices," *IEEE Transactions on Nuclear Science*, vol. 33, no. 6, pp. 1177–1184, 1986.
48. M. Shaneyfelt, P. Dodd, B. Draper, and R. Flores, "Challenges in Hardening Technologies Using Shallow-Trench Isolation," *IEEE Transactions on Nuclear Science*, vol. 45, no. 6, pp. 2584–2592, 1998.
49. G. Youk, P. Khare, R. Schrimpf, L. Massengill, and K. Galloway, "Radiation-Enhanced Short Channel Effects due to Multi-Dimensional Influence from Charge at Trench Isolation Oxides," *IEEE Transactions on Nuclear Science*, vol. 46, no. 6, pp. 1830–1835, 1999.
50. F. Faccio, S. Michelis, D. Cornale, A. Paccagnella, and S. Gerardin, "Radiation-Induced Short Channel (RISCE) and Narrow Channel (RINCE) Effects in 65 and 130 nm MOSFETs," *IEEE Transactions on Nuclear Science*, vol. 62, no. 6, pp. 2933–2940, 2015.
51. G. Grosso and G. P. Parravicini, *Solid State Physics*. Academic Press, 2013.
52. V. Ferlet-Cavrois, V. Pouget, D. McMorrow, J. R. Schwank, N. Fel, E. Essely, R. S. Flores, P. Paillet, M. Gaillardin, D. Kobayashi, J. S. Melinger, O. Duhamel, P. E. Dodd, and M. R. Shaneyfelt, "Investigation of the Propagation Induced Pulse Broadening (PIPB) Effect on Single Event Transients in SOI and Bulk Inverter Chains," *IEEE Transactions on Nuclear Science*, vol. 55, no. 6, pp. 2842–2853, 2008.
53. W. Morris, "Latchup in CMOS," in *2003 IEEE International Reliability Physics Symposium Proceedings, 2003. 41st Annual.*, 2003, pp. 76–84.
54. D. G. Mavis, D. R. Alexander, and G. L. Dinger, "A Chip-Level Modeling Approach for Rail Span Collapse and Survivability Analyses," *IEEE Transactions on Nuclear Science*, vol. 36, no. 6, pp. 2239–2246, 1989.
55. S. Veeraraghavan and J. Fossum, "Short-Channel Effects in SOI MOSFETs," *IEEE Transactions on Electron Devices*, vol. 36, no. 3, pp. 522–528, 1989.

56. S. Zhang, "Review of Modern Field Effect Transistor Technologies for Scaling," in *Journal of Physics: Conference Series*, vol. 1617, no. 1. IOP Publishing, 2020, p. 012054.
57. N. S. Saks, M. G. Ancona, and J. A. Modolo, "Radiation Effects in MOS Capacitors with Very Thin Oxides at 80°K," *IEEE Transactions on Nuclear Science*, vol. 31, no. 6, pp. 1249–1255, 1984.
58. D. M. Fleetwood, "Evolution of Total Ionizing Dose Effects in MOS Devices with Moore's Law Scaling," *IEEE Transactions on Nuclear Science*, vol. 65, no. 8, pp. 1465–1481, 2018.
59. H. Hughes and J. Benedetto, "Radiation Effects and Hardening of MOS Technology: Devices and Circuits," *IEEE Transactions on Nuclear Science*, vol. 50, no. 3, pp. 500–521, 2003.
60. J. Schwank, V. Ferlet-Cavrois, M. Shaneyfelt, P. Paillet, and P. Dodd, "Radiation Effects in SOI Technologies," *IEEE Transactions on Nuclear Science*, vol. 50, no. 3, pp. 522–538, 2003.
61. M. Mounir Mahmoud, J. Prinzie, D. Söderström, K. Niskanen, V. Pouget, A. Cathelin, S. Clerc, and P. Leroux, "Impact of Aging Degradation on Heavy-Ion SEU Response of 28-nm UTBB FD-SOI Technology," *IEEE Transactions on Nuclear Science*, vol. 69, no. 8, pp. 1865–1875, 2022.
62. M. Gaillardin, M. Raine, P. Paillet, M. Martinez, C. Marcandella, S. Girard, O. Duhamel, N. Richard, F. Andrieu, S. Barraud *et al.*, "Radiation Effects in Advanced SOI Devices: New Insights into Total Ionizing Dose and Single-Event Effects," in *2013 IEEE SOI-3D-Subthreshold Microelectronics Technology Unified Conference (S3S)*. IEEE, 2013, pp. 1–2.
63. M. Gaillardin, M. Martinez, P. Paillet, F. Andrieu, S. Girard, M. Raine, C. Marcandella, O. Duhamel, N. Richard, and O. Faynot, "Impact of SOI Substrate on the Radiation Response of UltraThin Transistors Down to the 20 nm Node," *IEEE Transactions on Nuclear Science*, vol. 60, no. 4, pp. 2583–2589, 2013.
64. X. Zhou, Z. Li, Z. Yuan, R. Wang, L. Shu, T. Wang, M. Qiao, Z. Wang, Z. Li, and B. Zhang, "Total-Ionizing-Dose Radiation-Induced Dual-Channel Leakage Current at Unclosed Edge Termination for High Voltage SOI LDMOS," *IEEE Transactions on Electron Devices*, vol. 68, no. 6, pp. 2861–2866, 2021.
65. P. E. Dodd, M. R. Shaneyfelt, J. R. Schwank, and J. A. Felix, "Current and Future Challenges in Radiation Effects on CMOS Electronics," *IEEE Transactions on Nuclear Science*, vol. 57, no. 4, pp. 1747–1763, 2010.
66. Z. Ren, X. An, G. Li, G. Chen, M. Li, G. Yu, Q. Guo, X. Zhang, and R. Huang, "TID Response of Bulk Si PMOS FinFETs: Bias, Fin Width, and Orientation Dependence," *IEEE Transactions on Nuclear Science*, vol. 67, no. 7, pp. 1320–1325, 2020.
67. T. Ma, S. Bonaldo, S. Mattiazzo, A. Baschirotto, C. Enz, A. Paccagnella, and S. Gerardin, "TID Degradation Mechanisms in 16-nm Bulk FinFETs Irradiated to Ultrahigh Doses," *IEEE Transactions on Nuclear Science*, vol. 68, no. 8, pp. 1571–1578, 2021.
68. H. Hughes, P. McMarr, M. Alles, E. Zhang, C. Arutt, B. Doris, D. Liu, R. Southwick, and P. Oldiges, "Total Ionizing Dose Radiation Effects on 14 nm FinFET and SOI UTBB Technologies," in *2015 IEEE Radiation Effects Data Workshop (REDW)*, 2015, pp. 1–6.
69. T. Liu, C. Geng, Z. Zhang, F. Zhao, S. Gu, T. Tong, K. Xi, G. Liu, Z. Han, M. Hou *et al.*, "Impact of Temperature on Single Event Upset Measurement by Heavy Ions in SRAM Devices," *Journal of Semiconductors*, vol. 35, no. 8, p. 084008, 2014.
70. D. Kobayashi, "Scaling Trends of Digital Single-Event Effects: A Survey of SEU and SET Parameters and Comparison With Transistor Performance," *IEEE Transactions on Nuclear Science*, vol. 68, no. 2, pp. 124–148, 2021.
71. K. LaBel, D. Hawkins, J. Kinnison, W. Stapor, and P. Marshall, "Single Event Effect Characteristics of CMOS Devices Employing Various Epi-Layer Thicknesses," in *Proceedings of the Third European Conference on Radiation and its Effects on Components and Systems*, 1995, pp. 258–262.

72. Z. Li, L. Berti, J. Wouters, J. Wang, and P. Leroux, "Characterization of the Total Charge and Time Duration for Single-Event Transient Voltage Pulses in a 65-nm CMOS Technology," *IEEE Transactions on Nuclear Science*, vol. 69, no. 7, pp. 1593–1601, 2022.
73. G. Torrens, S. A. Bota, B. Alorda, and J. Segura, "An Experimental Approach to Accurate Alpha-SER Modeling and Optimization Through Design Parameters in 6T SRAM Cells for Deep-Nanometer CMOS," *IEEE Transactions on Device and Materials Reliability*, vol. 14, no. 4, pp. 1013–1021, 2014.
74. C.-C. Liu, O. Lau, and J. Y. Du, "Complete DFM Model for High-Performance Computing SoCs with Guard Ring and Dummy Fill Effect," *ArXiv*, vol. abs/1701.00460, 2016.
75. J. Karp, M. J. Hart, P. Maillard, G. Hellings, and D. Linten, "Single-Event Latch-Up: Increased Sensitivity from Planar to FinFET," *IEEE Transactions on Nuclear Science*, vol. 65, no. 1, pp. 217–222, 2017.
76. R. Chen, F. Zhang, W. Chen, L. Ding, X. Guo, C. Shen, Y. Luo, W. Zhao, L. Zheng, H. Guo, Y. Liu, and D. M. Fleetwood, "Single-Event Multiple Transients in Conventional and Guard-Ring Hardened Inverter Chains Under Pulsed Laser and Heavy-Ion Irradiation," *IEEE Transactions on Nuclear Science*, vol. 64, no. 9, pp. 2511–2518, 2017.
77. G. S. Cardoso and T. R. Balen, "Study of Layout Extraction Accuracy on W/L Estimation of ELT in Analog Design Flow," in *2016 IEEE 7th Latin American Symposium on Circuits and Systems (LASCAS)*, 2016, pp. 279–282.
78. F. Faccio and G. Cervelli, "Radiation-Induced Edge Effects in Deep Submicron CMOS Transistors," *IEEE Transactions on Nuclear Science*, vol. 52, no. 6, pp. 2413–2420, 2005.
79. T. Calin, M. Nicolaidis, and R. Velazco, "Upset Hardened Memory Design for Submicron CMOS Technology," *IEEE Transactions on Nuclear Science*, vol. 43, no. 6, pp. 2874–2878, 1996.
80. H.-H. K. Lee, L. Klas, B. Mounaim, R. Prasanthi, I. R. Linscott, U. S. Inan, and M. Subhasish, "LEAP: Layout Design through Error-Aware Transistor Positioning for Soft-Error Resilient Sequential Cell Design," in *2010 IEEE International Reliability Physics Symposium*, 2010, pp. 203–212.
81. R. Ladbury, "Radiation Hardening at the System Level," in *IEEE NSREC Short Course*, 2007, pp. 1–94.
82. Assurance, Space Product, "Techniques for Radiation Effects Mitigation in AASIC and FPGAs Handbook," Technical report, ESA Requirements and Standards Division, Tech. Rep., 2016.
83. H.-Y. Jun and Y.-S. Lee, "Single Error Correction, Double Error Detection and Double Adjacent Error Correction with No Mis-Correction Code," *IEICE Electronics Express*, vol. 10, no. 20, p. 20130743, 2013.
84. J. M. Mogollón, F. R. Palomo, M. A. Aguirre, J. Nápoles, H. Guzmán-Miranda, and E. García-Sánchez, "TCAD Simulations on CMOS Propagation Induced Pulse Broadening Effect: Dependence Analysis on the Threshold Voltage," *IEEE Transactions on Nuclear Science*, vol. 57, no. 4, pp. 1908–1914, 2010.
85. F. Márquez, F. Munoz, L. Sanz, F. Palomo, and M. Aguirre, "AFTU, an Analog Single Event Effects Automatic Analysis Tool," in *Proc. 5th Int. Workshop Analogue and Mixed-Signal Integrated Circuits for Space Applications*, 2014.
86. F. Márquez, F. Munoz, F. Palomo, M. Aguirre, and M. Ullán, "Analysis of Single Event Transient Effects in Analogue Topologies," *Analogue and Mixed-Signal Integrated Circuits for Space Applications*, no. 1, 2012.
87. A. Nikolaou, M. Bucher, N. Makris, A. Papadopoulou, L. Chevas, G. Borghello, H. D. Koch, K. Kloukinas, T. S. Poikela, and F. Faccio, "Extending a 65nm CMOS Process Design Kit for High Total Ionizing Dose Effects," in *2018 7th International Conference on Modern Circuits and Systems Technologies (MOCAST)*, 2018, pp. 1–4.

88. S. Verhaegen, W. Sijbers, G. Pollissard, L. Bertil, G. Franciscatto, and G. Thys, "DARE SET Simulation Flow Integrated in Virtuoso ADE L/XL Design Environment," in *Proc. Int. Workshop Analogue Mixed-Signal Integr. Circuits Space Appl.(AMICSA)*, 2018, pp. 1–3.
89. G. Thys, S. Redant, E. Geukens, Y. Geerts, M. Fossion, and M. Melotte, "Radiation Hardened Mixed-signal IP with DARE technology," *Proc AMICSA*, 2012.
90. T. Rajkowski, F. Saigné, and P. Wang, "Radiation Qualification by Means of the System-Level Testing: Opportunities and Limitations," *Radiation Tolerant Electronics*, p. 123, 2022.
91. V. Girones, J. Boch, A. Carapelle, A. Chapon, T. Maraine, T. Labau, F. Saigné, and R. G. Alía, "The Use of High-Energy X-Ray Generators for TID Testing of Electronic Devices," *IEEE Transactions on Nuclear Science*, vol. 70, no. 8, pp. 1982–1989, 2023.
92. M. Rousselet, P. C. Adell, D. J. Sheldon, J. Boch, H. Schone, and F. Saigné, "Use and Benefits of COTS Board Level Testing for Radiation Hardness Assurance," in *2016 16th European Conference on Radiation and Its Effects on Components and Systems (RADECS)*, 2016, pp. 1–5.
93. C. Virmontois, J.-M. Belloir, M. Beaumel, A. Vriet, N. Perrot, C. Sellier, J. Bezine, D. Gambart, D. Blain, E. Garcia-Sanchez, W. Mouallem, and A. Bardoux, "Dose and Single-Event Effects on a Color CMOS Camera for Space Exploration," *IEEE Transactions on Nuclear Science*, vol. 66, no. 1, pp. 104–110, 2019.
94. P. O'Neill, G. Badhwar, and W. Culpepper, "Risk Assessment for Heavy Ions of Parts Tested with Protons," *IEEE Transactions on Nuclear Science*, vol. 44, no. 6, pp. 2311–2314, 1997.
95. S. M. Guertin, "Board Level Proton Testing Book of Knowledge for NASA Electronic Parts and Packaging Program," 2017.
96. A. de Bibikoff and P. Lamberbourg, "Method for System-Level Testing of COTS Electronic Board Under High-Energy Heavy Ions," *IEEE Transactions on Nuclear Science*, vol. 67, no. 10, pp. 2179–2187, 2020.
97. R. G. Alía, P. F. Martínez, M. Kastriotou, M. Brugger, J. Bernhard, M. Cecchetto, F. Cerutti, N. Charitonidis, S. Danzeca, L. Gatignon, A. Gerbershagen, S. Gilardoni, N. Kerboub, M. Tali, V. Wyrwoll, V. Ferlet-Cavrois, C. Boatella Polo, H. Evans, G. Furano, and R. Gaillard, "Ultra-energetic Heavy-Ion Beams in the CERN Accelerator Complex for Radiation Effects Testing," *IEEE Transactions on Nuclear Science*, vol. 66, no. 1, pp. 458–465, 2019.
98. P. Fernández-Martínez, R. G. Alía, M. Cecchetto, M. Kastriotou, N. Kerboub, M. Tali, V. Wyrwoll, M. Brugger, C. Cangialosi, F. Cerutti, S. Danzeca, M. Delrieux, R. Froeschl, L. Gatignon, S. Gilardoni, J. Lendaro, I. Mateu, F. Ravotti, H. Wilkens, and R. Gaillard, "SEE Tests With Ultra Energetic Xe Ion Beam in the CHARM Facility at CERN," *IEEE Transactions on Nuclear Science*, vol. 66, no. 7, pp. 1523–1531, 2019.
99. PULSCAN, "A Modular System for Laser Stimulation of Scaling Technologies," 2023. [Online]. Available: https://www.pulscan.com/pages/pulsys.php.
100. A. Coronetti, R. G. Alía, J. Budroweit, T. Rajkowski, I. D. Costa Lopes, K. Niskanen, D. Söderström, C. Cazzaniga, R. Ferraro, S. Danzeca, J. Mekki, F. Manni, D. Dangla, C. Virmontois, N. Kerboub, A. Koelpin, F. Saigné, P. Wang, V. Pouget, A. Touboul, A. Javanainen, H. Kettunen, and R. C. Germanicus, "Radiation Hardness Assurance Through System-Level Testing: Risk Acceptance, Facility Requirements, Test Methodology, and Data Exploitation," *IEEE Transactions on Nuclear Science*, vol. 68, no. 5, pp. 958–969, 2021.
101. J. Karp, M. J. Hart, P. Maillard, G. Hellings, and D. Linten, "Single-Event Latch-Up: Increased Sensitivity From Planar to FinFET," *IEEE Transactions on Nuclear Science*, vol. 65, no. 1, pp. 217–222, 2018.
102. S. Verhaegen, W. Sijbers, S. Zagrocki, L. Berti, J. Wouters, G. Franciscatto, G. Thys, S. Redant, B. Glass, and R. Jansen, "Incorporating More In-Depth Radiation Knowledge in the DARE180U Analog Design Kit," in *Proceeding of the Analog and Mixed Signal Integrated Circuits for Space Applications conference*, 2016, pp. 57–62.

References

103. M. S. Gorbunov, E. H. Boufouss, Z. Li, B. Vignon, M. D. v. d. Burgwal, L. Berti, and G. Thys, "Total ionizing dose effects sensitivity of unsalicided polysilicon resistors," *IEEE Transactions on Nuclear Science*, vol. 71, no. 8, pp. 1872–1878, 2024.
104. M. Glorieux, A. Evans, V. Ferlet-Cavrois, C. Boatella-Polo, D. Alexandrescu, S. Clerc, G. Gasiot, and P. Roche, "Detailed set measurement and characterization of a 65 nm bulk technology," *IEEE Transactions on Nuclear Science*, vol. 64, no. 1, pp. 81–88, 2017.
105. R. W. Blaine, S. E. Armstrong, J. S. Kauppila, N. M. Atkinson, B. D. Olson, W. T. Holman, and L. W. Massengill, "Rhbd bias circuits utilizing sensitive node active charge cancellation," *IEEE Transactions on Nuclear Science*, vol. 58, no. 6, pp. 3060–3066, 2011.
106. V. Ferlet-Cavrois, P. Paillet, D. McMorrow, N. Fel, J. Baggio, S. Girard, O. Duhamel, J. S. Melinger, M. Gaillardin, J. R. Schwank, P. E. Dodd, M. R. Shaneyfelt, and J. A. Felix, "New insights into single event transient propagation in chains of inverters–evidence for propagation-induced pulse broadening," *IEEE Transactions on Nuclear Science*, vol. 54, no. 6, pp. 2338–2346, 2007.
107. L. Chevas, A. Nikolaou, M. Bucher, N. Makris, A. Papadopoulou, A. Zografos, G. Borghello, H. D. Koch, and F. Faccio, "Investigation of scaling and temperature effects in total ionizing dose (tid) experiments in 65 nm cmos," in *2018 25th International Conference "Mixed Design of Integrated Circuits and System" (MIXDES)*, 2018, pp. 313–318.
108. L. T. Clark, C. S. Young-Sciortino, S. M. Guertin, W. E. Brown, K. E. Holbert, P. Bikkina, S. Bhanushali, A. Levy, M. Turowski, and J. D. Butler, "Extracting total ionizing dose threshold voltage shifts from ring oscillator circuit response," *IEEE Transactions on Device and Materials Reliability*, vol. 23, no. 1, pp. 162–171, 2023.
109. T. D. England, R. Arora, Z. E. Fleetwood, N. E. Lourenco, K. A. Moen, A. S. Cardoso, D. McMorrow, N. J.-H. Roche, J. H. Warner, S. P. Buchner, P. Paki, A. K. Sutton, G. Freeman, and J. D. Cressler, "An investigation of single event transient response in 45-nm and 32-nm soi rf-cmos devices and circuits," *IEEE Transactions on Nuclear Science*, vol. 60, no. 6, pp. 4405–4411, 2013.
110. P. Buckens and M. Veatch, "A high performance peak-detect and hold circuit for pulse height analysis," *IEEE Transactions on Nuclear Science*, vol. 39, no. 4, pp. 753–757, 1992.
111. P. Wang, A. L. Sternberg, J. A. Kozub, E. X. Zhang, N. A. Dodds, S. L. Jordan, D. M. Fleetwood, R. A. Reed, and R. D. Schrimpf, "Analysis of tpa pulsed-laser-induced single-event latchup sensitive-area," *IEEE Transactions on Nuclear Science*, vol. 65, no. 1, pp. 502–509, 2018.
112. V. Pouget, P. Fouillat, D. Lewis, H. Lapuyade, F. Darracq, and A. Touboul, "Laser cross section measurement for the evaluation of single-event effects in integrated circuits," *Microelectronics Reliability*, vol. 40, no. 8-10, pp. 1371–1375, 2000.
113. D. H. Habing, "The use of lasers to simulate radiation-induced transients in semiconductor devices and circuits," *IEEE Transactions on Nuclear Science*, vol. 12, no. 5, pp. 91–100, 1965.
114. F. K. Reed, *Radiation-hardened sensing and communication electronics with frequency drift correction using JFET technology*. Tennessee Technological University, 2022.
115. A. Karmakar, J. Wang, J. Prinzie, V. De Smedt, and P. Leroux, "A review of semiconductor based ionising radiation sensors used in harsh radiation environments and their applications," *Radiation*, vol. 1, no. 3, pp. 194–217, 2021.
116. J. Wang, J. Prinzie, A. Coronetti, S. Thys, R. G. Alia, and P. Leroux, "Study of seu sensitivity of sram-based radiation monitors in 65-nm cmos," *IEEE Transactions on Nuclear Science*, vol. 68, no. 5, pp. 913–920, 2021.
117. A. Virtanen, "The use of particle accelerators for space projects," in *Journal of Physics: conference series*, vol. 41, no. 1. IOP Publishing, 2006, p. 101.
118. Z. Li, L. Berti, Q. Lin, J. Zhao, M. Gorbunov, G. Thys, and P. Leroux, "An 80ms/s 70.79db-sndr 60.7fj/conv-step radiation-tolerant semi-time-interleaved pipelined-sar adc," in *2024 IEEE Custom Integrated Circuits Conference (CICC)*, 2024, pp. 1–2.

119. T. H. Kim and H. C. Lee, "Tid effect on a 12-bit 100ksps sar adc designed with a dummy gate-assisted n-mosfet," in *2015 IEEE Nuclear Science Symposium and Medical Imaging Conference (NSS/MIC)*, 2015, pp. 1–4.
120. B. Razavi, in *Design of Analog CMOS Integrated Circuits*. Boston, USA: McGraw-Hill Higher Education, 2000, ch. 12, pp. 439–443.
121. F. Wanlass and C. Sah, "Nanowatt Logic Using Field-Effect Metal-Oxide Semiconductor Triodes," in *1963 IEEE International Solid-State Circuits Conference. Digest of Technical Papers*, vol. VI, 1963, pp. 32–33.
122. H. Radamson and L. Thylén, *Monolithic Nanoscale Photonics-Electronics Integration in Silicon and Other Group IV Elements*. Academic Press, 2014.
123. Counterpoint, "Infographic: Global Foundry Revenue Share | Q2 2022," Sep 2022. [Online]. Available: https://www.counterpointresearch.com/insights/infographic-global-foundry-revenue-share-q2-2022/.
124. S. Williams, "Established Technology Nodes: The Most Popular Kid at the Dance," Sep 2016.
125. S. Leibson, "Costs for sub-20nm wafers put another nail in moore's law's coffin," Oct 2018.
126. L. Gonella, F. Faccio, M. Silvestri, S. Gerardin, D. Pantano, M. Re, M. Manghisoni, L. Ratti, and A. Ranieri, "Total Ionizing Dose Effects in 130-nm Commercial CMOS Technologies for HEP Experiments," *Nuclear Instruments and Methods in Physics Research Section A: Accelerators, Spectrometers, Detectors and Associated Equipment*, vol. 582, no. 3, pp. 750–754, 2007.
127. S. Bonacini, P. Valerio, R. Avramidou, R. Ballabriga, F. Faccio, K. Kloukinas, and A. Marchioro, "Characterization of a Commercial 65 nm CMOS Technology for SLHC Applications," *Journal of Instrumentation*, vol. 7, no. 01, p. P01015, 2012.
128. V. Ferlet-Cavrois, L. W. Massengill, and P. Gouker, "Single Event Transients in Digital CMOS–A Review," *IEEE Transactions on Nuclear Science*, vol. 60, no. 3, pp. 1767–1790, 2013.
129. V. Ferlet-Cavrois, P. Paillet, D. McMorrow, A. Torres, M. Gaillardin, J. Melinger, A. Knudson, A. Campbell, J. Schwank, G. Vizkelethy, M. Shaneyfelt, K. Hirose, O. Faynot, C. Jahan, and L. Tosti, "Direct Measurement of Transient Pulses Induced by Laser and Heavy Ion Irradiation in Deca-Nanometer Devices," *IEEE Transactions on Nuclear Science*, vol. 52, no. 6, pp. 2104–2113, 2005.
130. P. Eaton, J. Benedetto, D. Mavis, K. Avery, M. Sibley, M. Gadlage, and T. Turflinger, "Single Event Transient Pulsewidth Measurements Using a Variable Temporal Latch Technique," *IEEE Transactions on Nuclear Science*, vol. 51, no. 6, pp. 3365–3368, 2004.
131. J. A. Maharrey, J. S. Kauppila, R. C. Quinn, T. D. Loveless, E. X. Zhang, W. T. Holman, B. L. Bhuva, and L. W. Massengill, "Heavy-Ion Induced SETs in 32nm SOI Inverter Chains," in *2015 IEEE Radiation Effects Data Workshop (REDW)*, 2015, pp. 1–5.
132. B. Narasimham, V. Ramachandran, B. L. Bhuva, R. D. Schrimpf, A. F. Witulski, W. T. Holman, L. W. Massengill, J. D. Black, W. H. Robinson, and D. McMorrow, "On-Chip Characterization of Single-Event Transient Pulsewidths," *IEEE Transactions on Device and Materials Reliability*, vol. 6, no. 4, pp. 542–549, 2006.
133. T. Makino, D. Kobayashi, K. Hirose, Y. Yanagawa, H. Saito, H. Ikeda, D. Takahashi, S. Ishii, M. Kusano, S. Onoda, T. Hirao, and T. Ohshima, "LET Dependence of Single Event Transient Pulse-Widths in SOI Logic Cell," *IEEE Transactions on Nuclear Science*, vol. 56, no. 1, pp. 202–207, 2009.
134. T. D. Loveless, J. S. Kauppila, S. Jagannathan, D. R. Ball, J. D. Rowe, N. J. Gaspard, N. M. Atkinson, R. W. Blaine, T. R. Reece, J. R. Ahlbin, T. D. Haeffner, M. L. Alles, W. T. Holman, B. L. Bhuva, and L. W. Massengill, "On-Chip Measurement of Single-Event Transients in a 45 nm Silicon-on-Insulator Technology," *IEEE Transactions on Nuclear Science*, vol. 59, no. 6, pp. 2748–2755, 2012.

135. R. Harada, Y. Mitsuyama, M. Hashimoto, and T. Onoye, "Measurement Circuits for Acquiring SET Pulse Width Distribution with Sub-FO1-Inverter-Delay Resolution," *IEICE Transactions on Fundamentals of Electronics, Communications and Computer Sciences*, vol. 93, no. 12, pp. 2417–2423, 2010.
136. J. Chen, J. Yu, P. Yu, B. Liang, and Y. Chi, "Characterization of the Effect of Pulse Quenching on Single-Event Transients in 65-nm Twin-Well and Triple-Well CMOS Technologies," *IEEE Transactions on Device and Materials Reliability*, vol. 18, no. 1, pp. 12–17, 2018.
137. M. J. Gadlage, J. R. Ahlbin, B. L. Bhuva, N. C. Hooten, N. A. Dodds, R. A. Reed, L. W. Massengill, R. D. Schrimpf, and G. Vizkelethy, "Alpha-Particle and Focused-Ion-Beam-Induced Single-Event Transient Measurements in a Bulk 65-nm CMOS Technology," *IEEE Transactions on Nuclear Science*, vol. 58, no. 3, pp. 1093–1097, 2011.
138. Y. Chi, R. Song, S. Shi, B. Liu, L. Cai, C. Hu, and G. Guo, "Characterization of Single-Event Transient Pulse Broadening Effect in 65 nm Bulk Inverter Chains Using Heavy Ion Microbeam," *IEEE Transactions on Nuclear Science*, vol. 64, no. 1, pp. 119–124, 2017.
139. P. Dodd and L. Massengill, "Basic Mechanisms and Modeling of Single-Event Upset in Digital Microelectronics," *IEEE Transactions on Nuclear Science*, vol. 50, no. 3, pp. 583–602, 2003.
140. Cyclotron Resource Centre, UCLouvain. Parameters and Available Particles. [Online]. Available: https://uclouvain.be/en/research-institutes/irmp/crc/parameters-and-available-particles.html.
141. P. Dodd, O. Musseau, M. Shaneyfelt, F. Sexton, C. D'hose, G. Hash, M. Martinez, R. Loemker, J.-L. Leray, and P. Winokur, "Impact of Ion Energy on Single-Event Upset," *IEEE Transactions on Nuclear Science*, vol. 45, no. 6, pp. 2483–2491, 1998.
142. H. Puchner, D. Radaelli, and A. Chatila, "Alpha-Particle SEU Performance of SRAM with Triple Well," *IEEE Transactions on Nuclear Science*, vol. 51, no. 6, pp. 3525–3528, 2004.
143. P. Roche and G. Gasiot, "Impacts of Front-End and Middle-End Process Modifications on Terrestrial Soft Error Rate," *IEEE Transactions on Device and Materials Reliability*, vol. 5, no. 3, pp. 382–396, 2005.
144. G. Abadir, W. Fikry, H. Ragai, and O. Omar, "A Device Simulation and Model Verification of Single Event Transients in N/Sup +/-P Junctions," *IEEE Transactions on Nuclear Science*, vol. 52, no. 5, pp. 1518–1523, 2005.
145. G. Gasiot, D. Giot, and P. Roche, "Multiple Cell Upsets as the Key Contribution to the Total SER of 65 nm CMOS SRAMs and Its Dependence on Well Engineering," *IEEE Transactions on Nuclear Science*, vol. 54, no. 6, pp. 2468–2473, 2007.
146. T. Roy, A. F. Witulski, R. D. Schrimpf, M. L. Alles, and L. W. Massengill, "Single Event Mechanisms in 90 nm Triple-Well CMOS Devices," *IEEE Transactions on Nuclear Science*, vol. 55, no. 6, pp. 2948–2956, 2008.
147. S. DasGupta, A. F. Witulski, B. L. Bhuva, M. L. Alles, R. A. Reed, O. A. Amusan, J. R. Ahlbin, R. D. Schrimpf, and L. W. Massengill, "Effect of Well and Substrate Potential Modulation on Single Event Pulse Shape in Deep Submicron CMOS," *IEEE Transactions on Nuclear Science*, vol. 54, no. 6, pp. 2407–2412, 2007.
148. J. Zhang, J. Chen, P. Huang, S. Li, and L. Fang, "The Effect of Deep N+ Well on Single-Event Transient in 65 nm Triple-Well NMOSFET," *Symmetry*, vol. 11, no. 2, p. 154, 2019.
149. R. J. Van de Plassche, *Integrated analog-to-digital and digital-to-analog converters*. Springer Science & Business Media, 2012, vol. 264.
150. W. Kester, "A Brief History of Data Conversion: A Tale of Nozzles, Relays, Tubes, Transistors, and CMOS," *IEEE Solid-State Circuits Magazine*, vol. 7, no. 3, pp. 16–37, 2015.
151. H. Nyquist, "Certain Factors Affecting Telegraph Speed," *Transactions of the American Institute of Electrical Engineers*, vol. XLIII, pp. 412–422, 1924.
152. B. B. Monson, E. J. Hunter, A. J. Lotto, and B. H. Story, "The Perceptual Significance of High-Frequency Energy in the Human Voice," *Frontiers in psychology*, vol. 5, p. 587, 2014.

153. T. Sepke, P. Holloway, C. G. Sodini, and H.-S. Lee, "Noise Analysis for Comparator-Based Circuits," *IEEE Transactions on Circuits and Systems I: Regular Papers*, vol. 56, no. 3, pp. 541–553, 2009.
154. N. Bowers, "ADC Clock Input Considerations," Oct 2014. [Online]. Available: https://www.electronicspecifier.com/products/mixed-signal-analog/adc-clock-input-considerations.
155. R. J. Van de Plassche, *CMOS Integrated Analog-to-Digital and Digital-to-Analog Converters*. Springer Science & Business Media, 2013, vol. 742.
156. J. Fredenburg and M. P. Flynn, "ADC Trends and Impact on SAR ADC Architecture and Analysis," in *2015 IEEE Custom Integrated Circuits Conference (CICC)*. IEEE, 2015, pp. 1–8.
157. W. Kester, "ADC Architectures I: the Flash Converter," *Analog Devices Tutor. MT-020 Rev A*, 2009.
158. I. Ahmed, *Pipelined ADC Design and Enhancement Techniques*. Springer Science & Business Media, 2010.
159. P. J. A. Harpe, C. Zhou, Y. Bi, N. P. van der Meijs, X. Wang, K. Philips, G. Dolmans, and H. de Groot, "A 26 μ W 8 Bit 10 MS/s Asynchronous SAR ADC for Low Energy Radios," *IEEE Journal of Solid-State Circuits*, vol. 46, no. 7, pp. 1585–1595, 2011.
160. W. Kester, "Adc Architectures III: Sigma-Delta ADC Basics," *Analog Devices, MT022*, 2008.
161. P. Aziz, H. Sorensen, and J. vn der Spiegel, "An Overview of Sigma-Delta Converters," *IEEE Signal Processing Magazine*, vol. 13, no. 1, pp. 61–84, 1996.
162. C. C. Lee and M. P. Flynn, "A SAR-Assisted Two-Stage Pipeline ADC," *IEEE Journal of Solid-State Circuits*, vol. 46, no. 4, pp. 859–869, 2011.
163. L. Jie, X. Tang, J. Liu, L. Shen, S. Li, N. Sun, and M. P. Flynn, "An Overview of Noise-Shaping SAR ADC: From Fundamentals to the Frontier," *IEEE Open Journal of the Solid-State Circuits Society*, vol. 1, pp. 149–161, 2021.
164. L. Danial, N. Wainstein, S. Kraus, and S. Kvatinsky, "Breaking Through the Speed-Power-Accuracy Tradeoff in ADCs Using a Memristive Neuromorphic Architecture," *IEEE Transactions on Emerging Topics in Computational Intelligence*, vol. 2, no. 5, pp. 396–409, 2018.
165. B. Trump, "Radiation Handbook for Electronics," *Texas Instrum., Dallas, TX, USA, Tech. Rep*, 2023.
166. S. Christensson, I. Lundström, and C. Svensson, "Low frequency noise in mos transistors–i theory," *Solid-State Electronics*, vol. 11, no. 9, pp. 797–812, 1968.
167. T. Meisenheimer and D. Fleetwood, "Effect of radiation-induced charge on 1/f noise in mos devices," *IEEE Transactions on Nuclear Science*, vol. 37, no. 6, pp. 1696–1702, 1990.
168. H. J. Barnaby, "Total-ionizing-dose effects in modern cmos technologies," *IEEE Transactions on Nuclear Science*, vol. 53, no. 6, pp. 3103–3121, 2006.
169. F. Márquez, F. Muñoz, F. R. Palomo, L. Sanz, E. López-Morillo, M. A. Aguirre, and A. Jiménez, "Automatic Single Event Effects Sensitivity Analysis of a 13-Bit Successive Approximation ADC," *IEEE Transactions on Nuclear Science*, vol. 62, no. 4, pp. 1609–1616, 2015.
170. M. Andjelkovic, A. Ilic, Z. Stamenkovic, M. Krstic, and R. Kraemer, "An Overview of the Modeling and Simulation of the Single Event Transients at the Circuit Level," in *2017 IEEE 30th International Conference on Microelectronics (MIEL)*, 2017, pp. 35–44.
171. E. Pun-García and M. López-Vallejo, "A Survey of Analog-to-Digital Converters for Operation under Radiation Environments," *Electronics*, vol. 9, no. 10, p. 1694, 2020.
172. D. Zhang, A. Bhide, and A. Alvandpour, "A 53-nW 9.1-ENOB 1-kS/s SAR ADC in 0.13-μm CMOS for Medical Implant Devices," *IEEE Journal of Solid-State Circuits*, vol. 47, no. 7, pp. 1585–1593, 2012.
173. X. Tang, J. Liu, Y. Shen, S. Li, L. Shen, A. Sanyal, K. Ragab, and N. Sun, "Low-Power SAR ADC Design: Overview and Survey of State-of-the-Art Techniques," *IEEE Transactions on Circuits and Systems I: Regular Papers*, vol. 69, no. 6, pp. 2249–2262, 2022.

174. Y. Lim and M. P. Flynn, "A 1 mW 71.5 dB SNDR 50 MS/s 13 bit Fully Differential Ring Amplifier Based SAR-Assisted Pipeline ADC," *IEEE Journal of Solid-State Circuits*, vol. 50, no. 12, pp. 2901–2911, 2015.
175. H.-C. Hong and G.-M. Lee, "A 65-fJ/Conversion-Step 0.9-V 200-kS/s Rail-to-Rail 8-bit Successive Approximation ADC," *IEEE Journal of Solid-State Circuits*, vol. 42, no. 10, pp. 2161–2168, 2007.
176. M. Saberi, R. Lotfi, K. Mafinezhad, and W. A. Serdijn, "Analysis of Power Consumption and Linearity in Capacitive Digital-to-Analog Converters Used in Successive Approximation ADCs," *IEEE Transactions on Circuits and Systems I: Regular Papers*, vol. 58, no. 8, pp. 1736–1748, 2011.
177. C. Patel and C. Veena, "Study of Comparator and their Architectures," *International Journal of Multidisciplinary Consortium*, vol. 1, no. 1, pp. 1–12, 2014.
178. D. Zhang, C. Svensson, and A. Alvandpour, "Power Consumption Bounds for SAR ADCs," in *2011 20th European Conference on Circuit Theory and Design (ECCTD)*, 2011, pp. 556–559.
179. T. O. Anderson, "Optimum Control Logic for Successive Approximation Analog-to-Digital Converters," 1972. [Online]. Available: https://api.semanticscholar.org/CorpusID:59664024.
180. S. Hanfoug, N.-E. Bouguechal, and S. Barra, "Behavioral Non-Ideal Model of 8-Bit Current-Mode Successive Approximation Registers ADC by Using Simulink," *International Journal of u-and e-Service, Science and Technology*, vol. 7, no. 3, pp. 85–102, 2014.
181. X. Yu and J. Hu, "A New Low Leakage Power Flip-Flop Based on Ratioed Latches with Power Gating," *Procedia Environmental Sciences*, vol. 11, pp. 297–303, 2011.
182. Y.-K. Chang, C.-S. Wang, and C.-K. Wang, "A 8-bit 500-KS/s Low Power SAR ADC for Bio-medical Applications," in *2007 IEEE Asian Solid-State Circuits Conference*, 2007, pp. 228–231.
183. C.-C. Liu, S.-J. Chang, G.-Y. Huang, and Y.-Z. Lin, "A 10-bit 50-MS/s SAR ADC With a Monotonic Capacitor Switching Procedure," *IEEE Journal of Solid-State Circuits*, vol. 45, no. 4, pp. 731–740, 2010.
184. V. Hariprasath, J. Guerber, S. Lee, and U. Moon, "Merged Capacitor Switching Based SAR ADC with Highest Switching Energy-Efficiency," *Electronics letters*, vol. 46, no. 9, p. 620, 2010.
185. A. Sanyal and N. Sun, "SAR ADC Architecture with 98% Reduction in Switching Energy over Conventional Scheme," *Electronics Letters*, vol. 49, no. 4, pp. 248–250, 2013.
186. D. Zhu, R. Ding, and Z. Zhu, "99.8% Energy Saving and 97.4% Area Reduction Switching Scheme for SAR ADCs," *Analog Integrated Circuits and Signal Processing*, vol. 92, pp. 477–482, 2017.
187. W. Liu, T. Wei, B. Li, L. Yang, F. Xue, and Y. Hu, "A SAR-ADC Using Unit Bridge Capacitor and with Calibration for the Front-End Electronics of PET Imaging," *Nuclear Instruments and Methods in Physics Research Section A: Accelerators, Spectrometers, Detectors and Associated Equipment*, vol. 818, pp. 9–13, 2016.
188. Y. Chen, X. Zhu, H. Tamura, M. Kibune, Y. Tomita, T. Hamada, M. Yoshioka, K. Ishikawa, T. Takayama, J. Ogawa, S. Tsukamoto, and T. Kuroda, "Split Capacitor DAC Mismatch Calibration in Successive Approximation ADC," in *2009 IEEE Custom Integrated Circuits Conference*, 2009, pp. 279–282.
189. W. Liu, P. Huang, and Y. Chiu, "A 12-bit, 45-MS/s, 3-mW Redundant Successive-Approximation-Register Analog-to-Digital Converter With Digital Calibration," *IEEE Journal of Solid-State Circuits*, vol. 46, no. 11, pp. 2661–2672, 2011.
190. Y. Zhou, B. Xu, and Y. Chiu, "A 12 bit 160 MS/s Two-Step SAR ADC With Background Bit-Weight Calibration Using a Time-Domain Proximity Detector," *IEEE Journal of Solid-State Circuits*, vol. 50, no. 4, pp. 920–931, 2015.

191. J. A. McNeill, K. Y. Chan, M. C. W. Coln, C. L. David, and C. Brenneman, "All-Digital Background Calibration of a Successive Approximation ADC Using the Split ADC Architecture," *IEEE Transactions on Circuits and Systems I: Regular Papers*, vol. 58, no. 10, pp. 2355–2365, 2011.
192. J. McNeill, M. Coln, and B. Larivee, "Split ADC Architecture for Deterministic Digital Background Calibration of a 16-Bit 1-MS/s ADC," *IEEE Journal of Solid-State Circuits*, vol. 40, no. 12, pp. 2437–2445, 2005.
193. F. Ye, S. Li, M. Zhu, Z. Ni, and J. Ren, "A 13-bit 180-MS/s SAR ADC with Efficient Capacitor-Mismatch Estimation and Dither Enhancement," in *2019 IEEE International Symposium on Circuits and Systems (ISCAS)*, 2019, pp. 1–4.
194. M. Ding, P. Harpe, Y.-H. Liu, B. Busze, K. Philips, and H. de Groot, "A 46 μW 13 b 6.4 MS/s SAR ADC With Background Mismatch and Offset Calibration," *IEEE Journal of Solid-State Circuits*, vol. 52, no. 2, pp. 423–432, 2017.
195. Y. Zhu, C.-H. Chan, S.-S. Wong, U. Seng-Pan, and R. P. Martins, "Histogram-Based Ratio Mismatch Calibration for Bridge-DAC in 12-bit 120 MS/s SAR ADC," *IEEE Transactions on Very Large Scale Integration (VLSI) Systems*, vol. 24, no. 3, pp. 1203–1207, 2016.
196. H. Li, M. Maddox, M. C. W. Coin, W. Buckley, D. Hummerston, and N. Naeem, "A Signal-Independent Background-Calibrating 20b 1MS/S SAR ADC with 0.3ppm INL," in *2018 IEEE International Solid-State Circuits Conference - (ISSCC)*, 2018, pp. 242–244.
197. Z. Du, B. Yao, W. Xu, X. Wang, H. Hu, and L. Qiu, "Capacitor Mismatch Calibration of a 16-Bit SAR ADC Using Optimized Segmentation and Shuffling Scheme," *IEEE Transactions on Circuits and Systems II: Express Briefs*, vol. 70, no. 8, pp. 2789–2793, 2023.
198. Z. Li, L. Berti, G. Thys, and P. Leroux, "A cdac mismatch calibration technique for sar-assisted pipeline adcs," in *2023 21st IEEE Interregional NEWCAS Conference (NEWCAS)*, 2023, pp. 1–5.
199. C. J. González, B. L. Costa, D. N. Machado, R. G. Vaz, A. C. V. Bôas, O. L. Gonçalez, H. Puchner, F. L. Kastensmidt, N. H. Medina, M. A. Guazzelli et al., "Failure Mechanism and Sampling Frequency Dependency on TID Response of SAR ADCs," *Journal of Electronic Testing*, vol. 37, pp. 329–343, 2021.
200. B. L. Costa, C. J. González, R. G. Vaz, O. L. Gonçalez, and T. R. Balen, "Influence of Sampling Frequency on TID Response of SAR ADCs," in *2020 IEEE Latin-American Test Symposium (LATS)*, 2020, pp. 1–6.
201. S. İlik and M. B. Yelten, "Total Ionizing Dose (TID) Impact on Basic Amplifier Stages," *IEEE Transactions on Device and Materials Reliability*, vol. 23, no. 1, pp. 51–57, 2023.
202. P. R. Agostinho, O. L. Gonçalez, and G. Wirth, "Rail to Rail Radiation Hardened Operational Amplifier in Standard CMOS Technology with Standard Layout Techniques," *Microelectronics Reliability*, vol. 67, pp. 99–103, 2016.
203. P. Mróz, A. Otarola, T. A. Prince, R. Dekany, D. A. Duev, M. J. Graham, S. L. Groom, F. J. Masci, and M. S. Medford, "Impact of the SpaceX Starlink Satellites on the Zwicky Transient Facility Survey Observations," *The Astrophysical Journal Letters*, vol. 924, no. 2, p. L30, 2022.
204. M. Smith, D. Craig, N. Herrmann, E. Mahoney, J. Krezel, N. McIntyre, and K. Goodliff, "The Artemis Program: An Overview of NASA's Activities to Return Humans to the Moon," in *2020 IEEE Aerospace Conference*. IEEE, 2020, pp. 1–10.
205. H. Xu, Y. Cai, L. Du, Y. Zhou, B. Xu, D. Gong, J. Ye, and Y. Chiu, "28.6 A 78.5dB-SNDR Radiation- and Metastability-Tolerant Two-Step Split SAR ADC Operating up to 75MS/s with 24.9mW Power Consumption in 65nm CMOS," in *2017 IEEE International Solid-State Circuits Conference (ISSCC)*, 2017, pp. 476–477.
206. J. Kuppambatti, J. Ban, T. Andeen, P. Kinget, and G. Brooijmans, "A Radiation-Hard Dual Channel 4-Bit Pipeline for a 12-bit 40 MS/s ADC Prototype with Extended Dynamic Range

for the ATLAS Liquid Argon Calorimeter Readout Electronics Upgrade at the CERN LHC," *Journal of Instrumentation*, vol. 8, no. 09, p. P09008, 2013.
207. Z. Li, L. Berti, J. Zhao, Q. Lin, M. Gorbunov, S. Wang, G. Thys, and P. Leroux, "An 80 MS/s 70.8 dB-SNDR Radiation-Tolerant Semi-Time-Interleaved Pipelined-SAR ADC for Space Applications," *IEEE Transactions on Circuits and Systems I: Regular Papers*, pp. 1–14, 2025.
208. M. A. P. Pertijs and W. J. Kindt, "A 140 dB-CMRR Current-Feedback Instrumentation Amplifier Employing Ping-Pong Auto-Zeroing and Chopping," *IEEE Journal of Solid-State Circuits*, vol. 45, no. 10, pp. 2044–2056, 2010.
209. S. E. Armstrong, B. D. Olson, W. T. Holman, J. Warner, D. McMorrow, and L. W. Massengill, "Demonstration of a Differential Layout Solution for Improved ASET Tolerance in CMOS A/MS Circuits," *IEEE Transactions on Nuclear Science*, vol. 57, no. 6, pp. 3615–3619, 2010.
210. M. M. Mano, *Digital Design*. Pearson Educación, 2002.
211. J.-W. Nam and M. S.-W. Chen, "An Embedded Passive Gain Technique for Asynchronous SAR ADC Achieving 10.2 ENOB 1.36-mW at 95-MS/s in 65 nm CMOS," *IEEE Transactions on Circuits and Systems I: Regular Papers*, vol. 63, no. 10, pp. 1628–1638, 2016.
212. J.-W. Nam, D. Chiong, and M. S.-W. Chen, "A 95-MS/s 11-bit 1.36-mW Asynchronous SAR ADC with Embedded Passive Gain in 65nm CMOS," in *Proceedings of the IEEE 2013 Custom Integrated Circuits Conference*, 2013, pp. 1–4.
213. D. Xu, L. Qiu, Z. Zhang, T. Liu, L. Liu, K. Chen, and S. Xu, "A Linearity-Improved 8-bit 320-MS/s SAR ADC with Metastability Immunity Technique," *IEEE Transactions on Very Large Scale Integration (VLSI) Systems*, vol. 26, no. 8, pp. 1545–1553, 2018.
214. L. Chen, J. Ma, and N. Sun, "Capacitor Mismatch Calibration for SAR ADCs based on Comparator Metastability Detection," in *2014 IEEE International Symposium on Circuits and Systems (ISCAS)*, 2014, pp. 2357–2360.
215. H. Jeon and Y.-B. Kim, "A CMOS Low-Power Low-Offset and High-Speed Fully Dynamic Latched Comparator," in *23rd IEEE International SOC Conference*, 2010, pp. 285–288.
216. T. Lin, K.-S. Chong, W. Shu, N. K. Z. Lwin, J. Jiang, and J. S. Chang, "Experimental Investigation into Radiation-Hardening-by-Design (RHBD) Flip-Flop Designs in a 65nm CMOS Process," in *2016 IEEE International Symposium on Circuits and Systems (ISCAS)*, 2016, pp. 966–969.
217. N. A. Dodds, N. C. Hooten, R. A. Reed, R. D. Schrimpf, J. H. Warner, N. J.-H. Roche, D. McMorrow, S.-J. Wen, R. Wong, J. F. Salzman, S. Jordan, J. A. Pellish, C. J. Marshall, N. J. Gaspard, W. G. Bennett, E. X. Zhang, and B. L. Bhuva, "Effectiveness of SEL Hardening Strategies and the Latchup Domino Effect," *IEEE Transactions on Nuclear Science*, vol. 59, no. 6, pp. 2642–2650, 2012.
218. N.-C. Chen, P.-Y. Chou, H. Graeb, and M. P.-H. Lin, "High-Density MOM Capacitor Array with Novel Mortise-Tenon Structure for Low-Power SAR ADC," in *Design, Automation and Test in Europe Conference and Exhibition (DATE), 2017*, 2017, pp. 1757–1762.
219. K. Kobayashi, K. Kubota, M. Masuda, Y. Manzawa, J. Furuta, S. Kanda, and H. Onodera, "A Low-Power and Area-Efficient Radiation-Hard Redundant Flip-Flop, DICE ACFF, in a 65 nm Thin-BOX FD-SOI," *IEEE Transactions on Nuclear Science*, vol. 61, no. 4, pp. 1881–1888, 2014.
220. A. Watkins, K. Gnawali, and H. Quinn, "Single Event Effects Characterization Using a Single Photon Absorption Laser," in *2018 IEEE AUTOTESTCON*, 2018, pp. 1–4.
221. D. McMorrow, W. Lotshaw, J. Melinger, S. Buchner, and R. Pease, "Subbandgap Laser-Induced Single Event Effects: Carrier Generation via Two-Photon Absorption," *IEEE Transactions on Nuclear Science*, vol. 49, no. 6, pp. 3002–3008, 2002.
222. PULSCAN, "A Modular System for Laser Stimulation of Scaling Technologies," 2023. [Online]. Available: https://www.pulscan.com/pages/pulsys.php.

223. M. Muenchhof, M. Beck, and R. Isermann, "Fault-Tolerant Actuators and Drives–Structures, Fault Detection Principles and Applications," *Annual reviews in control*, vol. 33, no. 2, pp. 136–148, 2009.
224. L. Zhang-Li, H. Zhi-Yuan, Z. Zheng-Xuan, S. Hua, C. Ming, B. Da-Wei, N. Bing-Xu, and Z. Shi-Chang, "Bias Dependence of a Deep Submicron NMOSFET Response to Total Dose Irradiation," *Chinese Physics B*, vol. 20, no. 7, p. 070701, 2011.
225. Z. Li, A. Dutta, A. Mukherjee, X. Tang, L. Shen, L. He, and N. Sun, "A SAR ADC with Reduced kT/C Noise by Decoupling Noise PSD and BW," in *2020 IEEE Symposium on VLSI Circuits*. IEEE, 2020, pp. 1–2.
226. M.-J. Seo, Y.-D. Kim, J.-H. Chung, and S.-T. Ryu, "A 40nm CMOS 12b 200MS/s Single-Amplifier Dual-Residue Pipelined-SAR ADC," in *2019 Symposium on VLSI Circuits*. IEEE, 2019, pp. C72–C73.
227. "A 13-bit 160MS/s Pipelined Subranging-SAR ADC with Low-Offset Dynamic Comparator, author=Li, Weitao and Li, Fule and Liu, Jia and Li, Hongyu and Wang, Zhihua," in *2017 IEEE Asian Solid-State Circuits Conference (A-SSCC)*. IEEE, 2017, pp. 225–228.
228. Precedence Research, "Space Semiconductor Market," Oct 2023. [Online]. Available: https://www.precedenceresearch.com/space-semiconductor-market.
229. Statista, "Global Analog-to-Digital Converters Market Revenue 2018-2027," Oct 2021. [Online]. Available: https://www.statista.com/statistics/1058814/worldwide-analog-to-digital-converters-market-revenue.
230. Verifed Market Reports, "Rad-Hard Data Converter Market Size, Growth," Oct 2023. [Online]. Available: https://www.verifiedmarketreports.com/product/rad-hard-data-converter-market/.

MIX
Papier aus verantwortungsvollen Quellen
Paper from responsible sources
FSC® C105338

If you have any concerns about our products,
you can contact us on
ProductSafety@springernature.com

In case Publisher is established outside the EU,
the EU authorized representative is:
**Springer Nature Customer Service Center GmbH
Europaplatz 3, 69115 Heidelberg, Germany**

Printed by Libri Plureos GmbH
in Hamburg, Germany